ENVIRONMENTAL DECISIONS
IN THE FACE OF
UNCERTAINTY

Committee on Decision Making Under Uncertainty

Board on Population Health and Public Health Practice

INSTITUTE OF MEDICINE
OF THE NATIONAL ACADEMIES

THE NATIONAL ACADEMIES PRESS
Washington, D.C.
www.nap.edu

THE NATIONAL ACADEMIES PRESS   500 Fifth Street, NW   Washington, DC 20001

NOTICE: The project that is the subject of this report was approved by the Governing Board of the National Research Council, whose members are drawn from the councils of the National Academy of Sciences, the National Academy of Engineering, and the Institute of Medicine. The members of the committee responsible for the report were chosen for their special competences and with regard for appropriate balance.

This study was supported by Contract No. EP-C-09-003, TO#6 between the National Academy of Sciences and the Environmental Protection Agency. Any opinions, findings, conclusions, or recommendations expressed in this publication are those of the author(s) and do not necessarily reflect the view of the organizations or agencies that provided support for this project.

International Standard Book Number-13:   978-0-309-13034-9
International Standard Book Number-10:   0-309-13034-4

Additional copies of this report are available from the National Academies Press, 500 Fifth Street, NW, Keck 360, Washington, DC 20001; (800) 624-6242 or (202) 334-3313; http://www.nap.edu.

For more information about the Institute of Medicine, visit the IOM home page at: www.iom.edu.

Copyright 2013 by the National Academy of Sciences. All rights reserved.

Printed in the United States of America

The serpent has been a symbol of long life, healing, and knowledge among almost all cultures and religions since the beginning of recorded history. The serpent adopted as a logotype by the Institute of Medicine is a relief carving from ancient Greece, now held by the Staatliche Museen in Berlin.

Suggested citation: IOM (Institute of Medicine). 2013. *Environmental decisions in the face of uncertainty.* Washington, DC: The National Academies Press.

"Knowing is not enough; we must apply.
Willing is not enough; we must do."
—Goethe

INSTITUTE OF MEDICINE
OF THE NATIONAL ACADEMIES

Advising the Nation. Improving Health.

THE NATIONAL ACADEMIES
Advisers to the Nation on Science, Engineering, and Medicine

The **National Academy of Sciences** is a private, nonprofit, self-perpetuating society of distinguished scholars engaged in scientific and engineering research, dedicated to the furtherance of science and technology and to their use for the general welfare. Upon the authority of the charter granted to it by the Congress in 1863, the Academy has a mandate that requires it to advise the federal government on scientific and technical matters. Dr. Ralph J. Cicerone is president of the National Academy of Sciences.

The **National Academy of Engineering** was established in 1964, under the charter of the National Academy of Sciences, as a parallel organization of outstanding engineers. It is autonomous in its administration and in the selection of its members, sharing with the National Academy of Sciences the responsibility for advising the federal government. The National Academy of Engineering also sponsors engineering programs aimed at meeting national needs, encourages education and research, and recognizes the superior achievements of engineers. Dr. Charles M. Vest is president of the National Academy of Engineering.

The **Institute of Medicine** was established in 1970 by the National Academy of Sciences to secure the services of eminent members of appropriate professions in the examination of policy matters pertaining to the health of the public. The Institute acts under the responsibility given to the National Academy of Sciences by its congressional charter to be an adviser to the federal government and, upon its own initiative, to identify issues of medical care, research, and education. Dr. Harvey V. Fineberg is president of the Institute of Medicine.

The **National Research Council** was organized by the National Academy of Sciences in 1916 to associate the broad community of science and technology with the Academy's purposes of furthering knowledge and advising the federal government. Functioning in accordance with general policies determined by the Academy, the Council has become the principal operating agency of both the National Academy of Sciences and the National Academy of Engineering in providing services to the government, the public, and the scientific and engineering communities. The Council is administered jointly by both Academies and the Institute of Medicine. Dr. Ralph J. Cicerone and Dr. Charles M. Vest are chair and vice chair, respectively, of the National Research Council.

www.national-academies.org

COMMITTEE ON DECISION MAKING UNDER UNCERTAINTY

FRANK A. SLOAN (*Chair*), J. Alexander McMahon Professor of Health and Management and Professor of Economics, Duke University, Durham, NC
JAMES S. HOYTE, Assistant to the President and Associate Vice President, Adjunct Lecturer on Public Policy, Harvard University, Boston, MA (Retired)
ROGER E. KASPERSON, Research Professor and Distinguished Scientist, Clark University, Worcester, MA
EMMETT B. KEELER, Senior Mathematician and Professor of Health Economics, Pardee RAND Graduate School, Santa Monica, CA
SARAH B. KOTCHIAN, Associate Director for Planning, University of New Mexico, Albuquerque (Retired)
JOSEPH V. RODRICKS, Principal, ENVIRON International Corporation, Arlington, VA
SUSAN L. SANTOS, Assistant Professor, University of Medicine and Dentistry of New Jersey, Piscataway
STEPHEN H. SCHNEIDER,[1] Melvin and Joan Lane Professor for Interdisciplinary Environmental Studies, Department of Biology, and Senior Fellow, Woods Institute for the Environment, Stanford University, CA
STEPHANIE TAI, Assistant Professor of Law, University of Wisconsin Law School, Madison
DETLOF VON WINTERFELDT, Professor of Industrial and Systems Engineering and Professor of Public Policy and Management, University of Southern California, Los Angeles
ROBERT B. WALLACE, Irene Ensminger Steecher Professor of Epidemiology and Internal Medicine, University of Iowa, Iowa City

IOM Staff

MICHELLE C. CATLIN, Study Director (from September 2011)
KATHLEEN STRATTON, Study Director (through August 2011)
KRISTINA SHULKIN, Senior Project Assistant (until July 2008)
HOPE HARE, Administrative Assistant
ROSE MARIE MARTINEZ, Director, Board on Population Health and Public Health Practice

---

[1] Deceased, July 2010.

Reviewers

This report has been reviewed in draft form by persons chosen for their diverse perspectives and technical expertise, in accordance with procedures approved by the National Research Council's Report Review Committee. The purpose of this independent review is to provide candid and critical comments that will assist the institution in making its published report as sound as possible and to ensure that the report meets institutional standards for objectivity, evidence, and responsiveness to the study charge. The review comments and draft manuscript remain confidential to protect the integrity of the deliberative process. We wish to thank the following individuals for their review of this report:

Ann Bostrom, University of Washington
E. D. Elliott, Yale University School of Law
William H. Farland, Colorado State University
Adam M. Finkel, University of Medicine and Dentistry of New Jersey
Dennis G. Fryback, University of Wisconsin–Madison
Marianne Horinko, The Horinko Group
Ronald A. Howard, Stanford University
David O. Meltzer, University of Chicago
Kara Morgan, Food and Drug Administration
Richard D. Morgenstern, Resources for the Future
Mary D. Nichols, California Air Resources Board
Gregory M. Paoli, Risk Sciences International
Melissa J. Perry, George Washington University
David Spiegelhalter, Centre for Mathematical Sciences, Cambridge

Although the reviewers listed above have provided many constructive comments and suggestions, they were not asked to endorse the conclusions or recommendations nor did they see the final draft of the report before its release. The review of this report was overseen by **Chris G. Whipple,** ENVIRON, and **Harold C. Sox,** Geisel School of Medicine at Dartmouth. Appointed by the National Research Council and the Institute of Medicine, they were responsible for making certain that an independent examination of this report was carried out in accordance with institutional procedures and that all review comments were carefully considered. Responsibility for the final content of this report rests entirely with the authoring committee and the institution.

Contents

PREFACE	xvii
ABBREVIATIONS AND ACRONYMS	xix
SUMMARY	1

1 INTRODUCTION 19
Uncertainty and Environmental Decision Making, 20
The Context of This Report and the Charge to the Committee, 28
Committee Process, 30
Committee's Approach to the Charge, 31
Types of Uncertainty, 38
Overview of the Report, 43
References, 43

2 RISK ASSESSMENT AND UNCERTAINTY 47
Risk Assessment, 47
Uncertainty and Risk Assessment, 52
The History of Uncertainty Analysis, 56
Newer Approaches to Dealing with Uncertainties, 60
Examples of EPA's Risk Assessments, 61
Key Findings, 68
References, 69

3 UNCERTAINTY IN TECHNOLOGICAL AND ECONOMIC FACTORS IN EPA'S DECISION MAKING 73
Technology Availability, 73
Economics, 79
Other Factors, 99
Key Findings, 100
References, 102

4 UNCERTAINTY AND DECISION MAKING: LESSONS FROM OTHER PUBLIC HEALTH CONTEXTS 107
Uncertainty and Public Health Decisions, 108
Secondhand Smoke, 113
Listeria monocytogenes, 117
Bovine Spongiform Encephalopathy, 121
Contamination of the Food Supply with Melamine, 124
Avandia®, 127
Vaccination Decisions, 131
Lessons from History, 132
U.S. Preventive Services Task Force Recommendations, 137
Key Findings, 137
References, 139

5 INCORPORATING UNCERTAINTY INTO DECISION MAKING 147
Incorporating Uncertainty into a Decision-Making Framework, 149
Other Considerations, 171
Stakeholder Engagement, 173
Key Findings, 176
References, 178

6 COMMUNICATION OF UNCERTAINTY 181
Communication of Uncertainty in Risk Estimates, 182
Presentation of Uncertainty, 184
Considerations When Deciding on a Communications Approach, 198
Key Findings, 209
References, 211

7 SYNTHESIS AND RECOMMENDATIONS 217
Findings and Recommendations, 219
References, 226

APPENDIXES

A	Approaches to Accounting for Uncertainty	229
B	Committee Member Biographical Sketches	247
C	Meeting Agendas	255

Boxes, Figures, and Tables

BOXES

S-1 Committee's Statement of Task, 3
S-2 Implications of Uncertainty Analysis for Decision Making, 8

1-1 The Mission of the U.S. Environmental Protection Agency, 21
1-2 Committee's Statement of Task, 31
1-3 Uncertainty Versus Variability and Heterogeneity: The Committee's Use of the Terms, 39
1-4 An Example of a Decision in the Face of Deep Uncertainty, 42

2-1 Definitions, 48
2-2 Development of Estimates of Human Health Risks for Non-Cancer Endpoints, 49
2-3 Development of Estimates of Human Health Risks for Cancer Endpoints, 50
2-4 Trichloroethylene Risk Assessment: An Example of the Uncertainties Present in a Cancer Risk Assessment and How They Could Affect Regulatory Decisions, 54

3-1 EPA Control Technology Categories, 75
3-2 Definitions of Select Terms Used in Economic Analyses as Defined in *Guidelines for Preparing Economic Analysis*, 80

5-1 Definitions of Preliminary Graphical Representations of Uncertainty, 153
5-2 Implementing Value-of-Information in a Business Context, 167
5-3 Examples of Value-of-Information Measures, 169

6-1 Strengths and Weaknesses of Numeric, Verbal, and Visual Communication of Risk, 186
6-2 When Greater Attention to Reporting Uncertainties May Be Needed, 201

7-1 Implications of Uncertainty Analysis for Decision Making, 219

FIGURES

1-1 The Presidential/Congressional Commission on Risk Assessment and Risk Management's framework for risk management decisions, 26
1-2 Factors considered in EPA's decisions, 36

5-1a Framework for decision making, 148
5-1b Considerations for each assessment during phase 2, 149
5-2 Schematic illustrating the values that can be calculated in a value-of-information analysis, 168

6-1 Nine displays for communicating uncertain estimates for the value of a single variable used in experiments, 192
6-2 Examples of the most common graphical displays of uncertainty: (a) a probability density function, (b) a cumulative distribution function, and (c) a box-and-whisker plot, 194
6-3 Graphic used by Krupnick et al. (2006) to display sources of uncertainty and to describe the impact of each source of uncertainty on estimates of expected net benefits in 2025, 195

TABLES

1-1 Selected Statutory Requirements Related to Consideration of Factors Other Than Estimates of Human Health Risks, 34

4-1 Assessment of Risks, Benefits, Other Decision-Making Factors, and Uncertainty at Selected Public Health Agencies and Organizations, 109
4-2 Estimates of Infected Cases of BSE in the 20 Years Following Introduction of 500 Infected Animals into the United States, 122

5-1 Influence of the Type and Source of Uncertainty on Incorporating Uncertainty into a Decision, 154

6-1 Estimated Reduction in Nonfatal Acute Myocardial Infarctions Associated with Illustrative Attainment Strategies for the Revised and More Stringent Alternative PM NAAQS in 2020, 188
6-2 Supplemental Qualitative Table Used by the Intergovernmental Panel on Climate Change to Describe Its Confidence in Conclusions and Results, 190

Preface

Multiple sources of uncertainty exist in any risk assessment including those conducted by the U.S. Environmental Protection Agency (EPA), the lead agency responsible for protecting Americans against significant risks to human health and the environment. The EPA asked the Institute of Medicine (IOM) to convene a committee to provide guidance for its decision makers and partners on approaches to manage risk in different contexts when uncertainty is present. To tackle this issue, the IOM assembled a committee of experts in the fields of risk assessment, public health, health economics, decision analysis, public policy, risk communication, and environmental and public health law. The committee met five times, including three open sessions during which committee members discussed relevant issues with outside experts and discussed the charge with the EPA.

In discussing its charge, the committee found it helpful to clarify the questions in its statement of task. When considering that question of "how . . . uncertainty influence[s] risk management under different public health policy scenarios," the committee deliberated on how uncertainty can and should influence decisions and help decision makers, rather than focusing on how it currently influences such decisions. In addition, when considering tools and techniques from other areas of public health policy, the committee considered whether there are tools and techniques available from other decision-making settings of potential use to EPA decision making, what their benefits and drawbacks are, and whether and how those tools could be applied by EPA.

Uncertainty is a very broad topic with many potential implications for decision making; this presented a thorny challenge to the committee throughout its deliberations. That challenge was amplified by the broad range of perspectives and diverse backgrounds committee members brought to the deliberations. The result was adoption of a broader approach to considering uncertainty than is typically taken for environmental decisions. In contrast, historically, much of the work related to uncertainty by EPA and others has focused on the uncertainty in the estimates of human health risks.

Despite a lengthy delay in completing this report, and after responding to excellent peer-review comments, in the end, I am proud of the work we have done and hope that the EPA and other decision makers will find the fundamental report message useful. In summary, that message is that EPA has made substantial technical progress in how it conducts uncertainty analyses in support of its human health risk assessments. However, because uncertainties pervade not only relationships between hazards and health outcomes, more emphasis is needed on the uncertainty in factors affecting EPA's decisions in addition to estimates of uncertainties in how policies affect human health (e.g., uncertainty in economics and technological assessment that are used for regulatory purposes). Advances in accounting for these latter uncertainties are critical to more robust assessments and ultimately should lead to better decisions.

The committee would like to thank all of the individuals who contributed to the work of the committee, including those who presented to the committee (Appendix C), and the peer reviewers who gave the committee a careful assessment and a list of suggested changes that, when implemented, substantially improved the report. The committee also acknowledges the help of consultants Lynn Goldman and David Paltiel, who provided effective guidance at critical points in the Committee's work. I would also like to acknowledge committee members Michael Taylor and Robert Perciasepe, who resigned from the committee upon being offered appointments at the FDA and EPA, respectively, Dorothy Patton, who also resigned from the committee, and Steven Schneider, who died in July 2010. All four members made early contributions to the committee's deliberations but were not involved in the drafting and approval of the final report.

Finally, I would like to thank my colleagues on the committee for their efforts and perseverance throughout what turned out to be a lengthy process. They have argued their positions but also accommodated their colleagues and sought consensus. I would also like to acknowledge the contributions of a number of staff members from IOM, in particular Kathleen Stratton and Michelle Catlin, whose efforts were essential in information gathering, in report writing, in responding to reviewers' comments, and in providing the committee with assistance and support. Many thanks to many other IOM staff, particularly Rose Marie Martinez, who made important contributions along the way to the final production of this report.

Frank A. Sloan, *Chair*
Committee on Decision Making Under Uncertainty

Abbreviations and Acronyms

ACIP	Advisory Committee on Immunization Priorities
AHRQ	Agency for Healthcare Research and Quality
ANRF	American Nonsmokers' Rights Foundation
ATSDR	Agency for Toxic Substances and Disease Registry
BAT	best available technology economically achievable
B/C	benefit-to-cost
BCA	benefit–cost analysis
BCT	best conventional pollutant control technology
BMD	benchmark dose
BMR	benchmark response
BOD5	biochemical oxygen demand
BPT	best practicable control technology currently available
BSE	bovine spongiform encephalopathy
CAA	Clean Air Act
Cal EPA	California Environmental Protection Agency
CCSP	Climate Change Science Program
CDC	U.S. Centers for Disease Control and Prevention
CDER	Center for Drug Evaluation and Review
CDF	cumulative distribution function
CDRH	Center for Devices and Radiological Health
CEA	cost-effectiveness analysis
CERCLA	Comprehensive Environmental Response, Compensation, and Liability Act

CFSAN	Center for Food Safety and Nutrition
CGE	computable general equilibrium
CIN	cervical intraepithelial neoplasia
CJD	Creutzfeldt-Jakob disease
C–R	concentration–response
CWA	Clean Water Act
EIA	economic impact analysis
EPA	U.S. Environmental Protection Agency
EVIU	expected value of including uncertainty
EVPI	expected value of perfect information
EVSI	expected value of sample information
FAO	Food and Agriculture Organization of the United Nations
FDA	U.S. Food and Drug Administration
FSIS	Food Safety and Inspection Service
GACT	generally available control technology
GAO	U.S. Government Accountability Office
GRADE	Grading of Recommendations Assessment, Development and Evaluation
HAP	hazardous air pollutant
HCRA	Harvard Center for Risk Analysis
HHS	U.S. Department of Health and Human Services
HPV	human papillomavirus
IARC	International Agency for Research on Cancer
IOM	Institute of Medicine
IPCC	Intergovernmental Panel on Climate Change
LOAEL	lowest-observed-adverse-effect level
MACT	maximum achievable control technology
MCL	maximum containment level
MCLG	maximum containment level goal
MEI	maximally exposed individual
NAAQS	National Ambient Air Quality Standards
NAS	National Academy of Sciences
NESHAPS	National Emission Standards for Hazardous Air Pollutants
NOAEL	no-observed-adverse-effect level
NPV	net present value

NRC	National Research Council
NTP	National Toxicology Program
OMB	Office of Management and Budget
OSHA	Occupational Safety and Health Administration
OSTP	Office of Science and Technology Policy
PDF	probability density function
POD	point of departure
QALY	quality-adjusted life-year
RCRA	Resource Conservation and Recovery Act
RfC	reference concentration
RfD	reference dose
RTE	ready to eat
SAB	Science Advisory Board of the EPA
SDWA	Safe Drinking Water Act
SHS	secondhand smoke
SSRI	selective serotonin reuptake inhibitor
TCE	trichloroethylene
TDI	tolerable daily intake
TSE	transmissible spongiform encephalopathy
UF	uncertainty factor
USDA	U.S. Department of Agriculture
USPSTF	U.S. Preventive Services Task Force
VOI	value of information
VSL	value of statistical lives
WHO	World Health Organization
WTP	willingness to pay

Summary

The U.S. Environmental Protection Agency (EPA) is one of several federal agencies responsible for protecting Americans against significant risks to human health and the environment. As part of that mission, EPA estimates the nature, magnitude, and likelihood of risks to human health and the environment; identifies the potential regulatory actions that will mitigate those risks and protect public health[1] and the environment; and uses that information to decide on appropriate regulatory action. Uncertainties, both qualitative and quantitative, in the data and analyses on which these decisions are based enter into the process at each step. As a result, the informed identification and use of the uncertainties inherent in the process is an essential feature of environmental decision making.

MULTIPLE SOURCES OF UNCERTAINTY

This task is critical because of the multiple sources of uncertainty in the decision-making process. EPA has a long record of producing risk assessments and guidance documents relating to the analysis of uncertainty in estimating human health risks. Similarly, advisory bodies commenting on the role of uncertainty in EPA risk assessments and regulatory decisions have focused on the health risk component. However, EPA takes many other factors—economic and technological factors in particular—into

---

[1] Throughout this report the committee uses the term public health when referring to EPA's mission. The committee includes in the use of that term the whole population and individuals or individual subgroups within the whole population.

consideration when making its decisions, and the uncertainties in those other components are also worthy of attention. Unfortunately, the uncertainties in these areas receive much less attention than those in the area of human health, both from EPA and from advisory bodies. Social factors, such as environmental justice, and the political context also play a role in EPA's decisions and can have inherent uncertainties that are difficult to quantify.

This report strives to address this imbalance by giving attention to uncertainties in some of the factors that affect EPA's decision making in addition to the uncertainties in the estimates of human health risks. Although the committee distinguishes among the different factors in this report, the factors are not independent, and the lines between them are often blurred. Technological factors can affect an economic analysis in a number of ways. The cost of complying with a regulation might be estimated in a technological assessment, for example, but typically it would also be discussed as part of an economic assessment. The consideration of susceptible populations can affect estimates of health risks, and the socioeconomic status of a population affected by a regulation can affect estimates of a "willingness-to-pay" analysis conducted as part of an economic analysis. The political context can affect, explicitly or implicitly, the relative considerations given to the different factors in a decision.

This increasingly complex set of issues requires agreed-upon principles and analytical tools for conducting the uncertainty analyses used in making environmental decisions. As developed in this report, the use of those new tools in the analysis of uncertainty poses new challenges and opportunities for EPA in making and communicating its environmental decisions.

This summary opens with a description of EPA's charge to the committee and the committee's approach to the charge, followed by an overview of three types of uncertainty. Focusing next on the multiple sources of uncertainty and their use in decision making, the summary presents highlights from each section of the report. This summary closes with the committee's recommendations to EPA.

APPROACH TO THE CHARGE

Statement of Task

EPA requested that the Institute of Medicine convene a committee to provide guidance to its decision makers and their partners in states and localities on approaches to managing risk in different contexts when uncertainty is present. It also sought guidance on how information on uncertainty should be presented to help risk managers make sound decisions and to increase transparency in its communications with the public about

SUMMARY 3

> **BOX S-1**
> **Committee's Statement of Task**
>
> Based upon available literature, theory, and experience, the committee will provide its best judgment and rationale on how best to use quantitative information on the uncertainty in estimates of risk in order to manage environmental risks to human health and for communicating this information.
> Specifically, the committee will address the following questions:
>
> - How does uncertainty influence risk management under different public health policy scenarios?
> - What are promising tools and techniques from other areas of decision making on public health policy? What are benefits and drawbacks to these approaches for decision makers at EPA and their partners?
> - Are there other ways in which the EPA can benefit from quantitative characterization of uncertainty (e.g., value of information techniques to inform research priorities)?
> - What approaches for communicating uncertainty could be used to ensure the appropriate use of this risk information? Are there communication techniques to enhance the understanding of uncertainty among users of risk information like risk managers, journalists, and citizens?
> - What implementation challenges would EPA face in adopting these alternative approaches to decision making and communicating uncertainty? What steps should EPA take to address these challenges? Are there interim approaches that EPA could take?

those decisions. The specific questions that EPA requested the committee to address are presented in Box S-1.[2]

Given that its charge is not limited to human health risk assessment and includes broad questions about managing risks and decision making, in this report the committee examines the analysis of uncertainty in those other areas in addition to human health risks.

Types of Uncertainty

All EPA decisions involve uncertainty, but the type of uncertainty can vary widely from one decision to another. For an analysis of uncertainty to be useful or productive and for decision makers to determine when to invest

---

[2] Consistent with its charge, the committee focuses on "environmental risks to human health" in this report, and does not directly address ecological risk assessment. The committee notes, however, that many of the principles discussed and developed in this report would also apply to decision making related to ecological risks.

resources to reduce the uncertainty, a first and critical step is identifying the types of key uncertainties that contribute to a particular decision problem. The types of uncertainty also, in part, determine the best approaches for analyzing and communicating uncertainty. In this report, the committee classifies the various types of uncertainty into three categories: (1) statistical variability and heterogeneity (also called aleatory or exogenous uncertainty),[3] (2) model and parameter uncertainty (also called epistemic uncertainty), and (3) deep uncertainty (uncertainty about the fundamental processes or assumptions underlying a risk assessment).

Variability and *heterogeneity* refer to the natural variations in the environment, exposure paths, and susceptibility of subpopulations. They are inherent characteristics of a system under study, cannot be controlled by decision makers, and cannot be reduced by collecting more information. Empirical estimates of variability and heterogeneity can, however, be better understood through research in order to refine such estimates. Variability can often be quantified with standard statistical techniques, although it may be necessary to collect additional data.

Model[4] and parameter uncertainty include uncertainty due to the limited scientific knowledge about the nature of the models that link the causes and effects of environmental risks with risk-reduction actions as well as uncertainty about the specific parameters of the models. There may be various disagreements about the model, such as which model is most appropriate for the application at hand, which variables should be included in the model, the model's functional form (that is, whether the relationship being modeled is linear, exponential, or some other form), and how generalizable the findings based on data collected in another context are to the problem at hand (for example, the generalizability of findings based on experiments on animals to human populations). In theory, model and parameter uncertainty can be reduced by additional research.

Deep uncertainty is uncertainty that is not likely to be reduced by additional research within the time period in which a decision must be

---

[3] Although chemical risk assessors typically consider uncertainty and variability as separate and distinct, in the other areas uncertainty is seen as encompassing statistical variability and heterogeneity as well as model and parameter uncertainty. Because variability and heterogeneity can contribute to the uncertainty when a decision is being made, in this report the committee discusses them as a specific type of uncertainty.

[4] A "model" is defined in the National Research Council's *Science and Decisions: Advancing Risk Assessment* (2009, The National Academies Press) as a "simplification of reality that is constructed to gain insights into select attributes of a particular physical, biologic, economic, or social system. Mathematical models express the simplification in quantitative terms" (p. 96). Model parameters are "[t]erms in a model that determine the specific model form. For computational models, these terms are fixed during a model run or simulation, and they define the model output. They can be changed in different runs as a method of conducting sensitivity analysis or to achieve a calibration goal" (p. 97).

made. Typically, deep uncertainty is present when underlying environmental processes are not understood, when there is fundamental disagreement among scientists about the nature of the environmental processes, and when methods are not available to characterize the processes (such as the measurement and evaluation of chemical mixtures). When deep uncertainty is present, it is unclear how those disagreements can be resolved. In situations characterized by deep uncertainty, the probabilities associated with various regulatory options and associated utilities may not be known. Neither the collection and analysis of data nor expert elicitation to assess uncertainty is likely to be productive when key parties to a decision do not agree on the system model, prior probabilities, or the cost function. The task instead is to make decisions despite the presence of deep uncertainty using the available science and judgment, to communicate how those decisions were made, and to revisit those decisions when more information is available.

UNCERTAINTY IN EPA'S ESTIMATES OF HEALTH RISK

Uncertainty is inherent in the scientific information upon which health risk estimates are based. Uncertainties enter the health risk assessment process at every step and can be caused by the potential confounders in observational studies, by extrapolation from animal studies to human studies, by extrapolation from high- to low-dose exposures, by interindividual variability, and by modeling the relationships between concentrations, human exposures, and human health responses and evaluating the effect of interventions or risk control options on public health risk.

A number of reports from the National Research Council (NRC) and other bodies discuss the need to evaluate, assess, and communicate the uncertainties in such estimates. Many of those reports emphasize the need to quantify the uncertainties inherent in human health risk estimates, recommend moving away from the presentation of health risk as point estimates, detail the pitfalls of using defaults to capture uncertainties in those assessments, and urge the EPA to seek data that could supplant the use of defaults. To that end, EPA has been a leader in the development of quantitative approaches for uncertainty analysis, such as applying Monte Carlo analysis and Bayesian approaches to environmental risk assessments. These types of uncertainty analysis have been used in a broad variety of EPA risk assessments, ranging from complex analyses for major chemicals such as arsenic, methylmercury, and dioxin to work done for the multiplicity of chemicals entering the IRIS database or found at Superfund sites.

On the other hand, those analyses of and concerns about uncertainties have in some cases (such as in the agency's work involving dioxin contamination) delayed rulemaking. Furthermore, some uncertainty analyses have not provided useful or necessary information for the decision at hand.

Because of that, NRC and other organizations have cautioned against excessively complex uncertainty analysis and have emphasized the need for such analyses to be *decision driven*; that is, they have recommended that the amount of uncertainty analysis matches the needs of the decision maker. The connection of information to decision making is a key feature in value-of-information analyses.

This committee agrees that EPA often focuses on the analysis of uncertainty in human health risk estimates without considering the role of the uncertainty in the context of the decision, that is, without considering whether—or explaining how—the analysis influences the agency's regulatory decision. The magnitude of the uncertainty in risk estimates might not always be large enough to influence the decision, or the uncertainty in the estimates might be overshadowed by the uncertainty in the other factors that EPA considers in a decision (see below for discussion).

UNCERTAINTY IN COMPONENTS OF DECISION MAKING OTHER THAN ESTIMATES OF HUMAN HEALTH RISK

Data and analyses from fields other than human health risks play a role in EPA's decisions, including technological and economic considerations.[5] As with estimates of health risks, the three different types of uncertainty discussed above can be present, and the presence of uncertainty is usually unavoidable in those data and analyses. Some of EPA's technological feasibility and cost–benefit analyses assess uncertainty, but many do not. Furthermore, the contribution of uncertainties in other factors, such as social factors (for example, environmental justice) and the political context receive little or no attention. With the exception of EPA's new guidance on economic analysis, which includes a discussion of uncertainty analysis, the agency offers little guidance or information about how to assess uncertainty in factors other than health risks or about how it considers that uncertainty in its decisions.

UNCERTAINTY: OTHER PUBLIC HEALTH SETTINGS

Chapter 4 reviews the methods and processes used for uncertainty analysis at other public health agencies and organizations. Although a number of agencies conduct complex analyses that use probabilistic techniques to assess uncertainties in health risks, the tools and techniques that those

---

[5] Statutory requirements and constraints in the nation's environmental laws shape the overall decision-making process, with general requirements relating to data expectations, schedules and deadlines, public participation, and other considerations. At the same time these laws allow EPA considerable discretion in the development and implementation of environmental regulations.

organizations use are similar to those used by EPA. Thus, the committee did not identify promising tools and techniques for assessing uncertainty from other areas of public health that would present new guidance for EPA. There are some examples from those organizations, however, that illustrate important concepts in decision making in the face of uncertainty. For instance, an analysis of the effects of various regulatory options on the risks from bovine spongiform encephalopathy offers an example of an uncertainty analysis targeted to a regulatory decision (that is, a *decision-driven* assessment). An assessment of the risks of *Listeria monocytogenes* in different foods illustrates the importance of involving stakeholders and external experts early in the decision process in order to identify and decrease uncertainties; it also demonstrates how an assessment can help identify targeted risk-mitigation strategies. An assessment of the risks from melamine in infant formula shows how a simple risk assessment and uncertainty analysis can provide the information necessary to make a decision. FDA's handling of its Avandia® (rosiglitazone) decision illustrates how disagreements among scientists about scientific evidence can be communicated so that the public can understand the rationale for the ultimate decision. The lessons learned from an often-criticized decision about vaccinating during the 1976 pandemic scare emphasize the importance of a systematic approach to incorporating uncertainty into a decision and of an iterative approach to decision making under deep uncertainty.

INCORPORATING AND USING UNCERTAINTY IN DECISION MAKING

The appropriate uncertainty analysis for a decision—and how to consider uncertainty in a decision—will depend on the types, source, and magnitude of the uncertainty as well as on the context of the decision (for example, the severity of the adverse effects and the time frame within which a decision is needed). Although uncertainty analysis needs to be designed on a case-by-case basis, there are general frameworks and processes that should be followed for determining uncertainty analyses and how to incorporate uncertainty into a decision. The legal context of a decision will determine, in part, the degree of caution. In some contexts best estimates of risks will indicate the best action, whereas in others that require more caution, upper limits on risk will indicate the best action. In all cases, decision makers should explain why and how uncertainties were taken into account in their decisions. Systematically considering uncertainties and their potential to affect a decision from the onset of the decision-making process will improve the decision, focus uncertainty analyses on the decision at hand, facilitate the identification of uncertainties in factors in addition to health risk estimates, improve the planning of uncertainty analyses, and set the stage for the consideration of uncertainties in decisions. To that end it is important that uncertainty is considered and incorporated into each phase—problem formulation, risk assessment, and risk management—of a systematic decision-making process. Chapter 5 presents a framework for decision making that incorporates the planning, conducting, and considering of uncertainty analyses into the decision-making process.

The types and source of uncertainty are often key determinants of the appropriate type of uncertainty analysis. In Box S-2 the committee provides

---

BOX S-2
Implications of Uncertainty Analysis for Decision Making

Health Uncertainties
Uncertainty analyses in human health risk estimates can help decision makers to

- evaluate alternative regulatory options;
- assess how credible extreme risk estimates are and how much to rely on them in decision making;
- weigh the marginal decrease in risk against the effort made to reduce it;
- clarify issues within a decision by using scenarios to characterize very different worlds; and
- in the case of scenario analyses for deep uncertainty, identify regulatory solutions that are effective over a broad spectrum of scenarios.

Uncertainties About Technology Availability
Uncertainty analyses in technology availability can help decision makers to

- differentiate between well-established technologies with reasonably well-known costs, and those that have not been used for the purposes at hand; and
- consider which technology may be considered "best practicable" or "best available" by providing information about both the likelihood of success of the unproven technologies, the time frame for success, and the effectiveness if successful.

Uncertainties About Cost and Benefits
Given the highly uncertain estimates of both health benefits and costs, uncertainty analyses in cost–benefit analyses can inform decision makers about

- how difficult it is to differentiate among different potential decisions;
- the disagreement among experts about the way regulation affects the economy, even when using similar models; and
- the ranges and sensitivity of estimates to different variables.

guidance on how uncertainty analyses about health effects, technological availability, and cost can be used in the decison-making process. Decisions in the presence of deep uncertainty are particularly challenging. As discussed in Chapter 5, scenario analysis, value-of-information methods, and robust decision methods that allow for adaptive management can be useful under those circumstances.

There are no simple rules to translate uncertainty information into a decision. However, a decision maker should be informed about and appreciate the range of uncertainty when making a decision. How uncertainty analysis is used will differ depending on the type of uncertainty under consideration. Uncertainty analyses in human health risk estimates can help decision makers to weigh the marginal decrease in risk against the effort made to reduce it. Uncertainty analyses in technology availability can help decision makers to differentiate between well-established technologies with reasonably well-known costs and those that have not been used for the purposes at hand. Uncertainty analyses in cost–benefit analyses can inform decision makers about the disagreement among experts about the way regulations affect the economy, even when using similar models. Box S-2 provides other examples of the ways in which uncertainty analyses can be used in decision making.

The interpretation and incorporation of uncertainty into environmental decisions will depend on a number of characteristics of the risks and the decision. Those characteristics include the distribution of the risks, the decision makers' risk aversion, and the potential consequences of the decision.

The quality of the analysis and recommendations following from the analysis will depend on the relationship between the analyst and the decision maker. The planning, conduct, and results of uncertainty analysis should not be conducted in isolation, separated from the individuals who will eventually make the decisions. The success of a decision in the face of uncertainty depends on the analysts having a good understanding of the context of the decision and the information needed by the decision makers and also the decision makers having a good understanding of the evidence on which to base the decision, including understanding the uncertainty in that evidence.

COMMUNICATING UNCERTAINTY IN DECISIONS

Much of the research related to communicating uncertainty in environmental decisions actually focuses on communicating uncertainty in estimates of health risks. Research on risk communication, which includes the communication of the uncertainty in health risks, highlights the importance of using an interactive approach to communication. In other words, communication should not consist only of EPA providing information to others, but rather it should include an active exchange of information, with both EPA and stakeholders providing input to the conversation. Such interaction should occur from the onset of a decision—that is, during the problem-formulation phase of the decision-making process. Early communication facilitates incorporation of stakeholder perspectives in the process and helps to identify uncertainties for consideration in the decision. In addition, discussing known and potential uncertainties from the start of the decision-making process can increase social trust, which is critical for effective decision making, especially for decisions made in the face of high levels of uncertainty or scientific disagreement.

Uncertainty is typically expressed in terms of the probability or likelihood of an event and can be presented numerically, verbally, and graphically; each approach has its advantages and disadvantages. Numeric presentations can communicate a large amount of information but are useful only with audiences that are knowledgeable and capable of interpreting them. Verbal presentations (that is, the use of such terms as "likely" or "unlikely") can capture people's attention and portray directionality, but they can be prone to different interpretations in different contexts and by different people. Graphic presentations of probabilistic information can capture and hold people's attention, but individuals can vary in their ability to correctly interpret those presentations.

EPA communicates with people with a broad range of expertise and interests under a broad range of circumstances, and the appropriate communication approach will depend on who the communication is with, the source and types of the uncertainty, the context of the decision, and the purpose of the communication. For example, because the approach should be geared toward the audience, a table that includes numerical presentations of probabilities might be appropriate when communicating with a scientist or technical expert who is accustomed to thinking in those terms, but a graphical presentation might be more appropriate for a person without that background. Different people and groups of people, including scientists and regulatory decision makers, have biases (availability, confirmation, confidence, and group bias, among others) that can affect both the interpretation and the framing of uncertainty. Acknowledging those biases at the beginning of the decision-making process is critical to successful communication about an issue.

More communication is needed about the different sources of the uncertainty in EPA's decisions and about how those sources of uncertainty compare and affect a decision. For example, if the uncertainty in an estimate of health risks contributes to the uncertainty in a decision less than does the uncertainty in cost estimates, that should be communicated. Documenting the nature and magnitude of uncertainty in a decision is not only important at the time of the decision, but it is also important when a decision might be revisited or evaluated in the future.

FINDINGS AND RECOMMENDATIONS

Uncertainties in the Characterization of Human Health Risks

Finding 1

Decision documents[6] prepared to explain specific decisions often lack a robust discussion of the uncertainties identified in the health risk assessments prepared by agency scientists. Although those documents and communications should be succinct, open, and transparent, they should also include information on what uncertainties are present, which uncertainties need to be addressed, and how those uncertainties affected a decision. It should be clear from agency communications that **uncertainty is inherent in science, including the science that informs EPA decisions**. In addition to contributing to full transparency, providing information and fostering discussion of the existence of uncertainties, including unresolved uncertainty, could eventually lead to greater public understanding and appreciation of uncertainty in decision making.

RECOMMENDATION 1
To better inform the public and decision makers, U.S. Environmental Protection Agency (EPA) decision documents and other communications to the public should systematically

- include information on what uncertainties in the health risk assessment are present and which need to be addressed;
- discuss how the uncertainties affect the decision at hand; and
- include an explicit statement that uncertainty is inherent in science, including the science that informs EPA decisions.

Uncertainty in Other Factors That Influence a Decision

Finding 2

Although EPA decisions have included discussions and consideration of the uncertainties in the health risk assessment, the agency has generally given less attention to uncertainties in other contributors influencing the regulatory decision. Those contributors include economic and technological factors as well as other factors that are not easily quantified, such as environmental justice. A major challenge to decision making in the face of

---

[6] The committee uses the term "decision document" to refer to EPA documents that go from EPA staff to the decision maker and documents produced to announce an agency decision.

uncertainty is the uncertainty in those other factors. Methods and processes should be available for situations in which such analyses are appropriate and helpful to a decision maker. In general, this might require a research program to develop methods for this new type of uncertainty analysis, changes in decision documents and other analyses, and a program for research on communicating uncertainties.

RECOMMENDATION 2
The U.S. Environmental Protection Agency should develop methods to systematically describe and account for uncertainties in decision-relevant factors in addition to estimates of health risks—including technological and economic factors—in its decision-making process. When influential in a decision, those new methods should be subject to peer review.

Finding 3

EPA has developed guidance about, and conducted in-depth analyses of, the costs and benefits of major decisions. EPA guidance contains appropriate advice about the conduct of these analyses, including the discussion of some uncertainties. However, the committee noted a lack of transparency regarding uncertainty analyses in the cost–benefit assessments in some EPA decision documents. The information presented about these uncertainties is often arcane, hard to locate, and technically very challenging to non-experts. Those analyses often shape regulatory decisions; thus, they should be described in ways that are useful and interpretable for the decision maker and stakeholders. The needs of the two audiences—that is, technical and non-expert audiences—differ, but a given set of decision documents and supporting analyses could include descriptions that explain the sources of uncertainties to the non-expert and provide links, either electronically or via text, to more detailed descriptions of the economic analyses that are appropriate for experts.

RECOMMENDATION 3
Analysts and decision makers should describe in decision documents and other public communications uncertainties in cost–benefit analyses that are conducted, even if not required by statute for decision making, and the analyses should be described at levels that are appropriate for technical experts and non-experts.

Finding 4

The role of uncertainty in the costs and benefits and availability and feasibility of control technologies is not well investigated or understood. The evidence base for those factors is not robust. Evaluating case studies of past rulemaking and developing a directed research program on assessing the availability of technologies might be the first steps toward understanding the robustness of technology feasibility assessments and economic assessments as well as the potential for technology innovation.

RECOMMENDATION 4
The U.S. Environmental Protection Agency (EPA) should fund research, conduct research, or both to evaluate the accuracy and predictive capabilities of past assessments of technologies and costs and benefits for rulemaking in order to improve future efforts. This research could be conducted by EPA staff or else by nongovernmental policy analysts, who might be less subject to biases. This research should be used as a learning tool for EPA to improve its analytic approaches to assessing technological feasibility.

Finding 5

The committee did not find any specific guidance for assessing the uncertainties in the other factors that affect decision making, such as social factors (for example, environmental justice) and the political context. The committee also did not find examples of systematic consideration of those factors and their uncertainty when exploring the policy implications of strategies to mitigate harms to human health. In response to requirements in statutes or executive orders that require regulations to be based on the open exchange of information and the perspectives of stakeholders, some EPA programs (e.g., Superfund) work to address issues related to public (stakeholder) values and concerns.

Ecological risk assessments[7] have included contingent valuation to help inform policy development. Similarly, economists have explored the values people hold regarding specific health outcomes for the purposes of resource allocation or clinical guideline development. More research is needed into methods to appropriately characterize the uncertainty in those other factors and to communicate that uncertainty to decision makers and the public.

---

[7] Ecological risk assessment is a "process that evaluates the likelihood that adverse ecological effects may occur or are occurring as a result of exposure to one or more stressors" (http://www.epa.gov/raf/publications/pdfs/ECOTXTBX.PDF [accessed January 16, 2013]).

RECOMMENDATION 5
The U.S. Environmental Protection Agency should continue to work with stakeholders, particularly the general public, in efforts to identify their values and concerns in order to determine which uncertainties in other factors, along with those in the health risk assessment, should be analyzed, factored into the decision-making process, and communicated.

Finding 6

The nature of stakeholder participation in and input to a decision depends on the type of stakeholder. The regulated industry, local business communities, and environmental activists (including those at the local level, if they exist) are more likely to be proactively engaged in providing input on pending regulations. The general public, without encouragement or assistance from EPA (or local environmental regulatory departments), is less likely to participate effectively or at all in such activities. One means to bridge the gap in understanding the values of the public is a formal research program.

RECOMMENDATION 6
The U.S. Environmental Protection Agency should fund or conduct methodological research on ways to measure public values. This could allow decision makers to systematically assess and better explain the role that public sentiment and other factors that are difficult to quantify play in the decision-making process.

Framework for Incorporating Uncertainty in Decision Making

Finding 7

Uncertainty analysis must be designed on a case-by-case basis. The choice of uncertainty analysis depends on the context of the decision, including the nature or type of uncertainty, and the factors that are considered in the decision (that is, health risk, technological and economic factors, public sentiment, and the political context), as well as on the data that are available. Most environmental problems will require the use of multiple approaches to uncertainty analysis. As a result, a mix of statistical analyses and expert judgments will be needed.

A sensible, decision-driven, and resource-responsible approach to uncertainty analyses that includes decision makers and stakeholders is needed. Such a process will help ensure that the goals of the uncertainty analysis are consistent with the needs of the decision makers and the values and concerns of stakeholders, and it will help define analytic endpoints and identify population subgroups and heterogeneity and other uncertainties.

The committee believes that quantitative uncertainty analyses should only be undertaken when they are important and relevant to a given decision. Whether further quantitative uncertainty analysis is needed will depend on the ability of these analyses to affect the environmental decision at hand. One way to gauge this is to inquire whether perfect information would be able to change the decision, for example, whether knowing the exact dose–response function would change the regulatory regime. Clearly, if an environmental decision would stay the same for all states of information and analysis results, then it would not be worth conducting the analysis.

RECOMMENDATION 7
Although some analysis and description of uncertainty is always important, how many and what types of uncertainty analyses are carried out should depend on the specific decision problem at hand. The effort to analyze specific uncertainties through probabilistic risk assessment or quantitative uncertainty analysis should be guided by the ability of those analyses to affect the environmental decision.

Communication

Finding 8

A structured format for the public communication of the basis of EPA's decisions would facilitate transparency and subsequent work with stakeholders, particularly community members. EPA decision documents should make it clear that the identified uncertainties are in line with reasonable expectations presented in EPA guidelines and other sources. This practice would facilitate the goals of the first recommendation of the committee in this report—that *EPA decision documents should make it clear that uncertainty is inherent in agency risk assessments*. The committee intends that the recommendations in this report support full discussion of the difficulties of decision making, including—and possibly particularly—when social factors (such as environmental justice and public values) and political context play a large role.

RECOMMENDATION 8.1
U.S. Environmental Protection Agency (EPA) senior managers should be transparent in communicating the basis of the agency's decisions, including the extent to which uncertainty may have influenced decisions.

RECOMMENDATION 8.2
U.S. Environmental Protection Agency decision documents and communications to the public should include a discussion of which uncertainties are and are not reducible in the near term. The implications of each to policy making should be provided in other communication documents when it might be useful for readers.

Finding 9

Given that decision makers vary in their technical backgrounds and experience with highly mathematical depictions of uncertainty, a variety of communication tools should be developed. The public increasingly wants, and deserves, the opportunity to understand the decisions of appointed officials in order to manage their own risk and to hold decision makers accountable. With respect to *which* uncertainties or aspects of uncertainties to communicate, attention should be paid to the relevance to the audience of the uncertainties, so that the uncertainty information is meaningful to the decision-making process and the audience(s). Those efforts should include different types of decisions and should include communication of uncertainty to decision makers and to stakeholders and other interested parties.

RECOMMENDATION 9.1
The U.S. Environmental Protection Agency (EPA), alone or in collaboration with other relevant agencies, should fund or conduct research on communication of uncertainties for different types of decisions and to different audiences, develop a compilation of best practices, and systematically evaluate its communications.

RECOMMENDATION 9.2
As part of an initiative evaluating uncertainties in public sentiment and communication, U.S. Environmental Protection Agency senior managers should assess agency expertise in the social and behavioral sciences (for example, communication, decision analysis, and economics) and ensure it is adequate to implement the recommendations in this report.

In summary, the committee was impressed by the technical advances in uncertainty analysis used by EPA scientists in support of EPA's human health risk assessments, which form the foundation of all EPA decisions. The committee believes that EPA can lead the development of uncertainty analyses in economic and technological assessment that are used for regulatory purposes as well as the development of ways to characterize and account for public sentiment and political context. Leading in this way will require a targeted research program as well as disciplined attention to how those uncertainties are described and communicated to a variety of audiences, including the role that uncertainties have played in a decision.

1

Introduction

In an ideal world, scientists and the public would have exact, complete, and uncontested information on the factors responsible for pollution-related environmental and human health problems. Such information and analyses would allow decision makers to determine accurately and precisely which regulatory controls would lead to measurable benefits to human health and the environment, the costs associated with a regulatory control, and the time frame over which control measures would reduce exposure. Decision makers would be able to predict accurately whether contemplated regulations would lead to job loss or stimulate new programs, whether communities would be disrupted or energized, and whether regulation would unreasonably burden affected industries, state agencies, or tribes. In the absence of that information, decision makers would at least have an analysis of all uncertainties in the information and the expected impacts of those uncertainties on public health, the economy, and the public at large. Furthermore, in that ideal world time and resources would not be limiting, and the requisite information and analyses would be available at the time needed, and in the quality and quantity needed, so that decision makers would be able to make decisions consistently using relevant data.

In reality, however, decision makers do not have perfect information upon which to base decisions or to predict the impact and consequences of such decisions. But decisions need to be made despite those uncertainties. The available data on the multiple factors, including human health risk, that shape environmental decisions rarely encompass all relevant considerations. For example, although the toxic effects of a test substance may be definitive and undisputed in laboratory animals, the potential for risk to humans caused by environmental exposure to the same substance may be uncertain and controversial. Exposure levels for certain chemicals may be measurable and reproducible for populations in some locations but available only as modeled estimates in others. Investigators may have data to show that a pollutant is present in the soil at certain locations but may not have data to determine whether residents are exposed to it and, more importantly, whether residents are exposed at potentially harmful levels. Sophisticated methods for quantifying uncertainty may be available for some data categories but untested and of uncertain utility for others. Gathering additional information would require time and resources. As a result, uncertainty is always present in data and analysis, and decision making is invariably based on a combination of well-understood and less-well-understood information.

Consistent with its mission to "protect human health and the environment"[1] (Box 1-1) (EPA, 2011), the U.S. Environmental Protection Agency (EPA) estimates the nature, magnitude, and likelihood of risks to human health and the environment; identifies the potential regulatory actions for mitigating those risks and best protecting public health[2] and the environment; and considers the need to protect the public along with other factors when deciding on the appropriate regulatory action. Each of those steps has uncertainties that can be estimated qualitatively, quantitatively, or both qualitatively and quantitatively. A challenge for EPA is to determine how to best develop and use those estimates of uncertainty in data and analyses in its decision-making processes.

UNCERTAINTY AND ENVIRONMENTAL DECISION MAKING

Human health risk assessment[3] is one of the most powerful tools used by EPA in making regulatory decisions to manage threats to health and the environment. Historically, the analysis and consideration of uncertainty in decision making at EPA has focused on the uncertainties in the data and the analysis used in human health risk assessments, including the underlying sciences that comprise the field, such as exposure science, toxicology, and modeling.

---

[1] Many of the principles discussed in this report apply also to ecological risk assessment and decision making, but because the committee's charge focuses on human health, the report addresses issues relating only to the human health component of EPA's mission.

[2] Throughout this report the committee uses the term public health when referring to EPA's mission. The committee includes in the use of that term the whole population and individuals or individual subgroups within the whole population.

[3] Human health risk assessment is a systematic process within which scientific and other information relating to the nature and magnitude of threats to human health is organized and evaluated (NRC, 1983).

> **BOX 1-1**
> **The Mission of the U.S. Environmental Protection Agency**
>
> The mission of the U.S. Environmental Protection Agency (EPA) is to protect human health and the environment. Specifically, EPA's purpose is to ensure that
>
> - all Americans are protected from significant risks to human health and the environment where they live, learn, and work;
> - national efforts to reduce environmental risk are based on the best available scientific information;
> - federal laws protecting human health and the environment are enforced fairly and effectively;
> - environmental protection is an integral consideration in U.S. policies concerning natural resources, human health, economic growth, energy, transportation, agriculture, industry, and international trade, and these factors are similarly considered in establishing environmental policy;
> - all parts of society—communities, individuals, businesses, and state, local, and tribal governments—have access to accurate information sufficient to effectively participate in managing human health and environmental risks;
> - environmental protection contributes to making our communities and ecosystems diverse, sustainable, and economically productive; and
> - the United States plays a leadership role in working with other nations to protect the global environment.
>
> SOURCE: EPA, 2011.

The typical goal of a human health risk assessment is to develop a statement—the *risk characterization*—regarding the likelihood or probability that exposures arising from a given source, or in some cases from multiple sources, will harm human health. The risk characterization should include a statement about the scientific uncertainties associated with the assessment and their effect on the assessment, including a clear description of the confidence that the technical experts have in the results. Such information is provided to decision makers at EPA for consideration in their regulatory decisions (EPA, 2000; NRC, 2009). That information is also made available to the public. Uncertainties in the health risk assessment could stem, for example, from questions about how data from animals exposed to a chemical or other agent relate to human exposures or from uncertainties in the relationship between chemical exposures, especially low-dose chemical exposures, and a given adverse health outcome.

The statutes and processes guiding decision making at EPA make it clear that uncertainties in data and analyses are legitimate and predicable aspects of the decision-making process. Congress, the courts, and advisory bodies such as the National Research Council have recognized the inevitability of uncertainties in human health risk assessment and environmental decision making and, in some instances, have urged EPA to give special attention to this aspect of the process. The origins, necessity, and legitimacy of uncertainty as an aspect of EPA decision making, therefore, have both legal and scientific underpinnings. Those bases are discussed below.

Legal Basis

To fulfill its mission, EPA promulgates regulations to administer congressionally mandated programs. Although the statutes that govern EPA do not always contain explicit references to uncertainty in human health risk assessments, a number of statutes clearly imply that the information available to EPA may be uncertain and permit the agency to rely on uncertain information in its rulemaking. In other words, the need for EPA to consider and account for uncertainty when promulgating regulations is implicit in the statutes under which the agency operates. For example, the statutes related to air[4] and water[5] recognize and allow for uncertainty in decision making by calling for health and environmental standards with an "adequate" or "ample" margin of safety. Other statutes require judgments relating to the "potential" for environmental harm and "a reasonable certainty that no harm will result."[6] Congress's recognition of the uncertainty inherent in factors other than human health risks is also evident, for example, in such statements as "reasonably ascertainable economic consequences of the rule."[7] Such statements indicate a recognition by Congress that, at the time of a regulatory decision, data and information may be incomplete, controversial or otherwise open to variable interpretations, or that use of the available data and information may require assumptions about future events and conditions that are unknown or uncertain at the time of the rulemaking. Although the statutory language may seem vague and incomplete, the fact that such language was incorporated into a law by Congress indicates a recognition that EPA should have the discretion to interpret the statute and to develop approaches informed by agency experience, expertise, and decision-making needs.

---

[4] Clean Air Act Amendments of 1990, Pub. L. No. 101-549 (1990).
[5] Clean Water Act Amendments, Pub. L. No. 107-377, Sec. 1412(b)(3)(B)(iv) (1972).
[6] Federal Insecticide, Fungicide, and Rodenticide Act of 1947, Pub. L. No. 80-104 (1947).
[7] Toxic Substances Control Act of 1976, Pub. L. No. 94-469, Sec 6(c)(D) (1976).

Furthermore, some provisions do explicitly mandate that EPA discuss uncertainties in reports to Congress and other entities. For example, the Clean Air Act (CAA)[8] requires EPA to report to Congress on "any uncertainties in risk assessment methodology or other health assessment technique, and any negative health or environmental consequences to the community of efforts to reduce such risks." The Clean Water Act (CWA) Amendments[9] require EPA to specify, in a publicly available document, "to the extent practicable . . . each significant uncertainty identified in the process of the assessment of public health effects and studies that would assist in resolving the uncertainty."

In addition, several statutes contain provisions that amplify and clarify legislative objectives with respect to the uncertainties associated with assessing human health risk. For example, the 1996 Food Quality Protection Act (FQPA) specifies the following for pesticide approvals: "In the case of threshold effects . . . an additional ten-fold margin of safety for the pesticide chemical residues shall be applied for infants and children."[10] The 1996 amendments to the Safe Drinking Water Act (SDWA) are similarly explicit about the presentation of human health risk estimates and uncertainty in those estimates:

> The Administrator shall, in a document made available to the public in support of a regulation promulgated under this section, specify, to the extent practicable—
> (i) Each population addressed by any estimate of public health effects;
> (ii) The expected risk or central estimate of risk for the specific populations;
> (iii) Each appropriate upper-bound or lower-bound estimate of risk;
> (iv) Each significant uncertainty identified in the process of the assessment of public health effects and studies that would assist in resolving the uncertainty; and
> (v) Peer-reviewed studies known to the Administrator that support, are directly relevant to, or fail to support any estimate of public health effects and the methodology used to reconcile inconsistencies in the scientific data.[11]

Even from EPA's earliest days the courts have upheld the agency's legal authority and its need to account for uncertainty in its decision-making process. For example, in 1980 the U.S. Court of Appeals for the District of Columbia Circuit accepted and expanded on EPA's need to account for uncertainty by upholding an EPA decision related to the then new National Ambient Air Quality Standard (NAAQS) for particulate matter.[12]

---

[8] CAA Amendments of 1990, Pub. L. No. 101-549, Sec. 112(f)(1)(C) (1990).
[9] CWA Amendments, Pub. L. No. 107-377, Sec. 1412(b)(3)(B)(iv) (1972).
[10] 42 U.S.C. § 346a(b)(2)(C) (1996).
[11] SDWA Amendments of 1996, Pub. L. No. 104-182, § 300g-1(b)(3) (1996).
[12] 40 CFR pt. 50.

In affirming EPA's regulation under the CAA, the court pointed to "the Act's precautionary and preventive orientation"[13] and noted that "some uncertainty about the health effects of air pollution is inevitable"[14] and that "Congress provided that the Administrator is to use his judgment in setting air quality standards precisely to permit him to act in the face of uncertainty."[15]

Scientific Basis

As a discipline, science treats uncertainty as a natural and legitimate part of measurement methodology and, therefore, as an expected aspect of technical data used in decision making. To make scientific progress, "uncertainty, discrepancy, and inconsistency" are often necessary to point the way to new lines of experimentation and new discoveries (Lindley, 2006). However, the variability in the state of understanding and uncertainty is quite different from the more binary and absolute world of regulatory and courtroom decisions surrounding the EPA's regulatory decisions. To try to bridge that gap, the scientific community has endeavored to provide regulatory decision makers with a more comprehensive view of the estimates of risks, and a number of scientific reports have been published that focus on regulatory decision making and, to some extent, on the uncertainties inherent in the information that supports EPA's human health regulatory decisions and on the implications of that uncertainty.

Over the past three decades two core themes, which were outlined in the germinal report *Risk Assessment in the Federal Government: Managing the Process* (hereafter the Red Book) (NRC, 1983), have governed core aspects of human health risk assessment and decision making at EPA: (1) the special meaning of "uncertainty" in relation to human health risk assessment and decision making; and (2) the interface between risk assessment and risk management, that is, the interface between estimating risks and the decision about how to manage those risks. When discussing uncertainties, the Red Book (NRC, 1983) focused on uncertainty in risk estimates and, to some extent, in economic analyses, stating that "there is often great *uncertainty in estimates of the types, probability, and magnitude of health effects associated with a chemical agent, of the economic effects of a proposed regulatory action, and of the extent of current and possible future human exposures*" (p. 11; emphasis added). The report highlights

---

[13] Lead Industries Ass'n, Inc. v. EPA, 647 F.2d 1130, 1155 (D.C.Cir. 1980), 64 [http://openjurist.org/647/f2d/1130/lead-industries-association-inc-v-environmental-protection-agency].

[14] Lead Industries Ass'n, Inc. v. EPA, 647 F.2d 1130, 1155 (D.C.Cir. 1980), 63 [http://openjurist.org/647/f2d/1130/lead-industries-association-inc-v-environmental-protection-agency].

[15] Lead Industries Ass'n, Inc. v. EPA, 647 F.2d 1130, 1155 (D.C.Cir. 1980), 63 [http://openjurist.org/647/f2d/1130/lead-industries-association-inc-v-environmental-protection-agency].

the need to include uncertainties in characterizations of health risks, stating, "The summary effects of the *uncertainties in the preceding steps* are described in this step" (p. 20; emphasis added). It also highlights the need to communicate uncertainty and the paucity of guidance on how to do so, stating, "The final expressions of risk derived in [the risk characterization] will be used by the regulatory decision maker. . . . Little guidance is available on how to express *uncertainties in the underlying data* and on which dose–response assessments and exposure estimates should be combined to give a final estimate of possible risks" (p. 36; emphasis added).[16]

Over a decade later, *Science and Judgment in Risk Assessment: Managing the Process* (NRC, 1994), in response to its charge, focused on estimates of human health risks and statistical methods to quantify the uncertainty in those estimates. In line with those reports, EPA has focused a great deal of attention on methods for quantifying and expressing the uncertainty in health risk estimates.

Understanding Risk: Informing Decisions in a Democratic Society (hereafter *Understanding Risk*) (NRC, 1996), discussed the uncertainty inherent in estimates of health risks in the broad context of regulatory decisions, including those made by EPA. That committee wrote, "Significant advances have been made in recent years in the development of analytical methods for evaluating, characterizing, and presenting uncertainty and for analyzing its components, and well documented guidance for conducting an uncertainty analysis is available" (p. 108). The committee focused, however, "on the role of uncertainty in risk characterization and the role that uncertainty analysis can play as part of an effective iterative process for assessing, deliberating, and understanding risks when discussing uncertainties" (p. 108). It also proposed that "[p]erhaps the most important need is to identify and focus on uncertainties that matter to understanding risk situations and making decisions about them," and emphasized "the critical importance of social, cultural, and institutional factors in determining how uncertainties are considered, addressed, or ignored in the tasks that support risk characterization" (p. 108). In other words, *Understanding Risk* highlighted the subjective nature of interpreting uncertainty in human health risk estimates and how that subjectivity—which is influenced by social and cultural factors such as public values and preconceptions—can affect how the uncertainty is characterized.

---

[16] Responding to this emphasis in that and subsequent NRC reports, EPA treats risk characterization as a fundamental element in risk assessment and, importantly, the place where uncertainties analyzed in the course of the assessment are collected and described for decision makers. For this reason, risk characterization in EPA guidance documents almost always incorporates uncertainties. Similarly, "transparency" almost always embraces the idea of full disclosure of uncertainties, and the rationale for options considered and choices made, among other things.

The final report of the Presidential/Congressional Commission on Risk Assessment and Risk Management (1997a) recommended a risk management framework geared toward environmental risk decisions (see Figure 1-1). The three main principles of the framework are (1) putting health

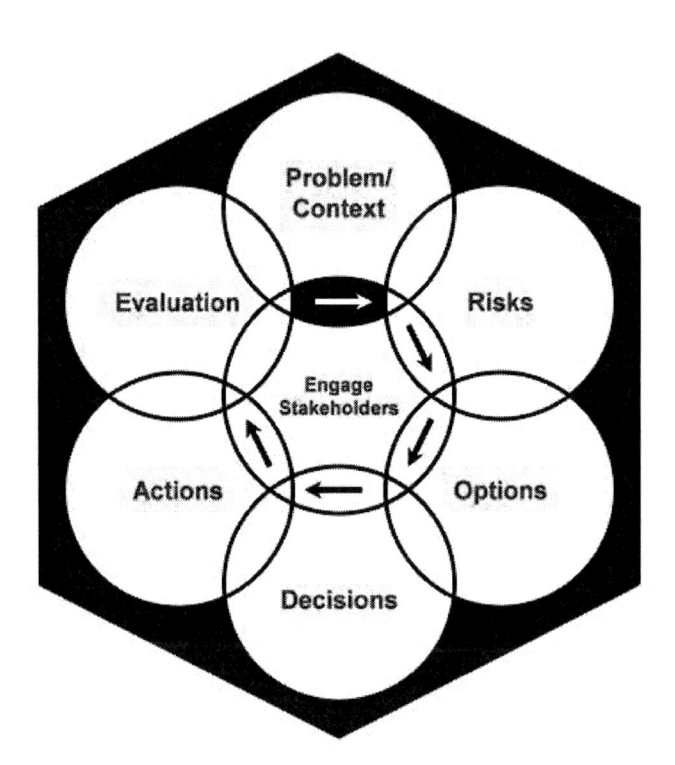

FIGURE 1-1 The Presidential/Congressional Commission on Risk Assessment and Risk Management's framework for risk management decisions. The commission designed the framework "to help all types of risk managers—government officials, private-sector businesses, individual members of the public—make good risk management decisions." The framework has six stages: define the problem and put it in context; analyze the risks associated with the problem in context; examine options for addressing the risks; make decisions about which options to implement; take actions to implement the decisions; and conduct an evaluation of the actions' results. The framework should be conducted in collaboration with stakeholders and should use "iterations if new information is developed that changes the need for or nature of risk management."
SOURCE: Presidential/Congressional Commission on Risk Assessment and Risk Management, 1997, p. 3.

and environmental problems in their larger, real-world contexts,[17] (2) involving stakeholders "to the extent appropriate and feasible during all stages of the risk management process" (p. 6), and (3) providing risk managers (referred to as the decision makers in this report) and stakeholders with opportunities to revisit stages within the framework when new information emerges. When discussing uncertainty in health risk characterizations for routine risk assessments, the report recommended using qualitative descriptions of uncertainty rather than quantitative analyses, because it "is likely to be more understandable and useful than quantitative estimates or models to [decision makers] and the public" (p. 170). In Volume 2 of its report, the commission further discussed uncertainty analyses, focusing on the uncertainties in both health risk estimates and economic analyses (Presidential/Congressional Commission on Risk Assessment and Risk Management, 1997b). Studies have shown, however, that different people interpret qualitative descriptions differently (Budescu et al., 2009; Wallsten and Budescu, 1995; Wallsten et al., 1986). See Chapter 6 for further discussion.

Science and Decisions: Advancing Risk Assessment (NRC, 2009) emphasized that risk assessment is an applied science used to help evaluate risk management options, and, as such, assessments should be conducted with that purpose in mind. It further stated that "descriptions of the uncertainty and variability inherent in all risk assessments may be complex or relatively simple; the level of detail in the descriptions should align with what is needed to inform risk-management decisions" (p. 5). The report recommended that EPA adopt a three-phase framework and employ a "broad-based discussion of risk-management options early in the process, extensive stakeholder participation throughout the process, and consideration of life-cycle approaches in a broader array of agency programs" (p. 260). Phase III of the framework, the risk-management phase, includes identifying the factors other than human health risk estimates that affect and are affected by the regulatory decision. These include the factors discussed above that EPA is required by certain statutes to consider, such as technologies and costs. The report does not, however, discuss the uncertainties in those factors and how any such uncertainties should affect EPA's decisions.

More recently, *A Risk-Characterization Framework for Decision Making at the Food and Drug Administration* (NRC, 2011) emphasized that "risk characterization should be decision-focused" and describe "potential

---

[17] "Evaluating problems in context involves evaluating different sources of a particular chemical or chemical exposure, considering other chemicals that could affect a particular risk or pose additional risks, assessing other similar risks, and evaluating the extent to which different exposures contribute to a particular health effect of concern" (Presidential/Congressional Commission on Risk Assessment and Risk Management, 1997b, p. 5).

effects of alternative *decisions* on health rather than on comparing different health and environmental hazards" (p. 21). The report discussed the factors that, in addition to human health risks, are sometimes considered in the decisions of the Food and Drug Administration (FDA): social factors, political factors, and economic factors.[18] Those factors often are not independent from the estimates of human health risks. The report briefly discussed uncertainty, focusing on the uncertainty in the characterization of risk, not the uncertainty in the other factors that play a role in FDA's decision making.

Over the past several decades EPA's science advisory boards (SABs) have also addressed the importance of considering uncertainties in risk assessment and decision making. For example, the SABs have emphasized its importance in radiological assessments (EPA, 1993, 1999b), the CAA (EPA, 2007), microbial risk assessments (EPA, 2010b), expert elicitation (EPA, 2010b), and a comparative risk framework methodology (EPA, 1999a).

These reports have contributed greatly to the science of risk assessment, uncertainty analysis, and environmental regulatory decision making. Although a number of those reports discuss the factors beyond the estimates of health risks that play a role in regulatory decision making, when discussing the analysis and implications of uncertainty on decision making they focus on the uncertainty in the estimates of human health risks. The reports typically do not discuss the uncertainty inherent in the other factors that are considered in regulatory decisions. This report broadens the discussion of uncertainty to include the uncertainty in factors in addition to human health risk assessments.

THE CONTEXT OF THIS REPORT AND THE CHARGE TO THE COMMITTEE

The major environmental statutes administered by EPA call for the agency to develop and use scientific data and analyses to evaluate potential human health risks. (See, for example, sections 108 and 109 of the CAA,[19] Section 3 of the Federal Insecticide, Fungicide, and Rodenticide Act

---

[18] EPA defines economic factors as the factors that "inform the manager on the cost of risks and the benefits of reducing them, the costs of risk mitigation or remediation options and the distributional effects" (EPA, 2000, p. 52). When discussing economic analysis, EPA includes the examination of net social benefits, impacts on industry, governments, and nonprofit organizations, and effects on various subpopulations, particularly low-income and minority individuals and children, using distributional analyses (EPA, 2010a).

[19] CAA of 1963, Pub. L. No. 2-206 (1963).

[FIFRA],[20] and Section 4 of the Toxic Substances Control Act [TSCA].[21]) As a result, the uncertainties inherent in scientific information—that is, in the data and analyses—are incorporated in the health risk assessment process. While a substantial database may exist for any particular chemical undergoing regulatory review, uncertainties invariably raise questions about the reliability of risk estimates and the scientific credibility of related regulatory decisions. Because some aspects of interpreting uncertainty are subjective, different risk assessors, regulators, and observers who approach the use of risk assessments and uncertainty analyses from different perspectives might have different interpretations of the results.

The presence of uncertainty has delayed decisions because it was thought or argued that more definitive answers to outstanding questions were needed and that research could provide those answers. Unfortunately, however, research cannot always provide more definitive information. It might not be possible at all to obtain that information, or obtaining it might require more time or resources than are available. In many cases additional research may address some forms of uncertainty while, at the same time, identifying other, new areas of uncertainty. Such delays can threaten public health by delaying the implementation of protective measures. The need for decision making despite uncertainty is highlighted in these passages from *Estimating the Public Health Benefits of Proposed Air Pollution Regulations*: "Even great uncertainty does not imply that action to promote or protect public health should be delayed," and "The potential for improving decisions through research must be balanced against the public health costs because of a delay in the implementation of controls. Complete certainty is an unattainable ideal" (NRC, 2002, p. 127). Delaying risk assessments because of uncertainty also diminishes the established practices that provide for a scientifically based, consistent approach for assessors to fill in gaps when information about a given chemical is lacking—for example, by extrapolating from animal data to humans and from high to low doses, by using uncertainty factors, by developing exposure scenarios, or by relying on other defaults. Delaying decisions also undervalues the estimating of uncertainties using simple qualitative or complex quantitative analyses.

There is an increasingly complex set of tools at EPA's disposal for uncertainty analyses in risk assessments. Those tools pose new challenges for decision making and for the communication of EPA's decisions, including the need to make deliberate decisions about when to use different tools or approaches and how best to communicate complex statistical analyses to nontechnical stakeholders. Rather than barring decisions in the face of

---

[20] FIFRA of 1947 to regulate the marketing of economic poisons and devices and for other purposes, Pub. L. No. 80-104 (1947).
[21] TSCA, Pub. L. No. 94-469 (1976).

uncertainty, transparency and scientific rigor have been used to help ensure the responsible use of uncertain information. The focus, however, remains on the uncertainty in the human health risk assessment, often with the uncertainties inherent in the other factors that play a role in EPA's decisions being ignored.

Given the challenges in decision making in the face of uncertainty, EPA requested that the Institute of Medicine (IOM) convene a committee to provide guidance to decision makers at EPA—and their partners in the states and localities—on approaches to managing risk in different contexts when confronting uncertainty. EPA also sought guidance on how information on uncertainty should be presented in order to best help decision makers make sound decisions and to increase transparency with the public about those decisions.

Specifically, EPA directed: "Based upon available literature, theory, and experience, the committee will . . . provide its best judgment and rationale on how best to use quantitative information on uncertainty in the estimates of risk in order to manage environmental risks to human health and for communicating this information."[22] The specific questions that EPA requested the committee to address are presented in Box 1-2.

COMMITTEE PROCESS

The IOM convened a committee of experts[23] in the fields of risk assessment, public health, health economics, decision analysis, public policy, risk communication, and environmental and public health law (see Appendix B for biographical sketches of committee members). The committee met five times, including two open sessions during which committee members discussed relevant issues with outside experts, and discussed the charge with the EPA. Appendix C presents the agendas of the open sessions.

---

[22] Consistent with its charge, the committee focused on "environmental risks to human health" in this report and did not directly address ecological risk assessment. The committee notes, however, that many of the principles discussed and developed in this report would also apply to decision making related to ecological risks.

[23] Michael Taylor and Robert Perciasepe resigned from the committee upon consideration of an appointment at the Food and Drug Administration and EPA, respectively. Steven Schneider died in July 2010. All three members contributed a great deal to the intellectual development of the report but were not involved in the approval of the final report.

> **BOX 1-2**
> **Committee's Statement of Task**
>
> Based upon available literature, theory, and experience, the committee will provide its best judgment and rationale on how best to use quantitative information on the uncertainty in estimates of risk in order to manage environmental risks to human health and for communicating this information.
> Specifically, the committee will address the following questions:
>
> - How does uncertainty influence risk management under different public health policy scenarios?
> - What are promising tools and techniques from other areas of decision making on public health policy? What are benefits and drawbacks to these approaches for decision makers at EPA and their partners?
> - Are there other ways in which EPA can benefit from quantitative characterization of uncertainty (e.g., value of information techniques to inform research priorities)?
> - What approaches for communicating uncertainty could be used to ensure the appropriate use of this risk information? Are there communication techniques to enhance the understanding of uncertainty among users of risk information like risk managers, journalists, and citizens?
> - What implementation challenges would EPA face in adopting these alternative approaches to decision making and communicating uncertainty? What steps should EPA take to address these challenges? Are there interim approaches that EPA could take?

COMMITTEE'S APPROACH TO THE CHARGE

Audience for the Report

The committee considered decision makers at EPA to be a main audience for this report. In addition, the committee prepared this report for a broader audience, which includes other environmental professionals, journalists, and interested observers.

Questions Within the Statement of Task

The committee found it helpful to clarify the questions within its statement of task. The committee was asked "how . . . uncertainty influence[s] risk management under different public health policy scenarios?" However, the committee found it more useful to deliberate on how uncertainty *can* and *should* influence decisions and help decision makers, rather than focusing on how it currently influences risk-management decisions.

Although there is an emphasis in the charge on "how best to use quantitative information on the uncertainty" and how "EPA can benefit from quantitative characterization of uncertainty," other aspects of the charge necessitate a broader look at uncertainty analysis beyond quantitative analysis. For example, questions about how "uncertainty influence[s] risk management under different public health policy scenarios," the "tools and techniques from other areas of decision making," and "what approaches for communicating uncertainty could be used to ensure the appropriate use of this risk information" require looking at both qualitative and quantitative analyses of uncertainty. Descriptive and qualitative narratives about uncertainty are instructive for decision makers and the public and, often, are more readily available and more comprehensible than the more technical quantitative analyses. In fact, a descriptive narrative, or else using both descriptive and quantitative approaches, is sometimes more appropriate. Accordingly, the committee interprets the charge as calling for an examination of decision-making uncertainties in both descriptive and quantitative, or narrative and numerical, terms.

Sources of Uncertainty: Factors That Play a Role in Decision Making and Their Uncertainty

Decision making at EPA is a multifaceted process that involves a broad variety of laws, activities, participants, and products. Congressionally enacted statutes and executive orders establish the fundamental principles and primary components for consideration in EPA's decisions (see Table 1-1 for examples). While health impacts are an important component of all EPA decisions, some statutes require that a decision be based solely on health impacts without consideration of cost or other factors, whereas other statutes require a consideration of technological feasibility, costs, or both. For example, the drinking water standards promulgated under the Safe Drinking Water Act[24] require that certain costs and technological availability be considered. In contrast, the standards that EPA promulgates under the NAAQS Act[25] consider only the protection of public health, although state regulators can consider costs, feasibility, and other factors when developing implementation plans to meet the NAAQS.

In addition to statutory requirements, executive orders require government agencies, including EPA, to conduct benefit–cost analyses (BCAs) as part of a regulatory analysis when promulgating certain regulations unless otherwise prohibited by statute. (See below for further discussion of BCA.)

---

[24] SWDA, Pub. L. No. 93-523.
[25] 42 U.S.C. § 7409(b) (2010).

A series of executive orders (see, for example, Executive Orders 12866[26] and 13563[27] for executive agencies and 13579 for independent regulatory agencies) discuss the need for regulatory impact analyses, stating that agencies "should assess all costs and benefits of available regulatory alternatives, including the alternative of not regulating."[28] In 2003 the Office of Management and Budget (OMB) issued Circular A-4 which provides further guidance to agencies conducting such analyses. Executive Order 13563, in addition to discussing cost–benefit analyses, also requires agencies, including EPA, to develop a plan for conducting periodic, retrospective analyses of their regulations.[29] OMB also publishes a yearly report to Congress on the benefits and costs of regulations and mandates (OMB, 2011). That document summarizes the costs and benefits of agencies across the government and, to reflect some of the uncertainty, includes a range of potential benefits.

Executive orders also require a consideration of the effects of regulations on certain populations. For example, Executive Order 13045 requires certain analyses for rules that "concern an environmental health risk or safety risk that an agency has reason to believe may disproportionately affect children." Executive Order 12898, *Federal Actions to Address Environmental Justice in Minority Populations and Low-Income Populations*,[30] encourages agencies to conduct their activities "in a manner that ensures that such programs, policies, and activities do not have the effect of excluding persons (including populations) from participation in, denying persons (including populations) the benefits of, or subjecting persons (including populations) to discrimination under, such programs, policies, and activities, because of their race, color, or national origin."[31]

Factors that are involved in regulatory decision making have been discussed before. For example, EPA's *Risk Characterization Handbook* describes seven factors that influence decision making: risk assessment, economic factors, technological factors, legal factors, social factors, political factors, and public values (EPA, 2000). More recently, *A Risk-Characterization Framework for Decision Making at the Food and Drug Administration* presented a framework for decision makers at FDA that includes four factors considered in FDA's decisions: factors related to the estimates of risks to public health (that is, the risk-characterization factors), economic factors, social factors, and political factors (NRC, 2011). In Figure 1-2, this committee has modified that framework to illustrate the three main factors that, depending on the legal context, EPA considers in

---

[26] Exec. Order No. 12866. 58 FR 51735 (October 4, 1993).
[27] Exec. Order No. 13563. 76 FR 3821 (January 21, 2011).
[28] Exec. Order No. 13045. 62 FR 19885 (April 23, 1997).
[29] Exec. Order No. 13563. 76 FR 3821 (January 21, 2011).
[30] Exec. Order No. 12898. 77 FR 11752 (February 28, 2012).
[31] Exec. Order No. 12898 Section 2-2. 77 FR 11752 (February 28, 2012).

TABLE 1-1 Selected Statutory Requirements Related to Consideration of Factors Other Than Estimates of Human Health Risks

Statute and Program	Considerations Under the Statute
Clean Air Act—National Emission Standards for Hazardous Air Pollutants	To determine whether to add or delete a pollutant from the Hazardous Air Pollutants (HAPs) list, EPA must consider whether the "substance may reasonably be anticipated to cause any adverse effects to the human health or adverse environmental effects" but not costs or technical feasibility.[a] Costs are permissible considerations in EPA's determination of maximum achievable control technologies (MACTs) and generally available control technologies (GACTs). For MACTs, EPA can consider whether the control technology achieves the "maximum degree of reduction in emissions" of the HAPs as well as "the cost of achieving such emission reduction, and any non-air quality health and environmental impacts and energy requirements."[b] For GACTs, EPA may instead elect to use "generally available control technologies or management practices."[c]
National Ambient Air Quality Standards (NAAQS)	NAAQS themselves must be set to levels "requisite to protect the public health" "allowing for an adequate margin of safety" (for primary standards),[d] and levels "requisite to protect the public welfare from any known or anticipated adverse effects associated with the presence of such air pollutant in the ambient air" (for secondary standards).[e] The statute assigns implementation responsibilities to state regulators, who may consider costs, feasibility, and other factors in developing the required state implementation plans (SIPs).[f] In an area that does not achieve the NAAQS for a given pollutant, a proposed new source of that pollutant must achieve the lowest achievable emission rate[g] in order to receive the necessary permit. Thus with MACT and GACT determinations, uncertainties in determining costs as well as uncertainties in environmental effects may be relevant to a decision.

many of its decisions: (1) estimates of human health risks, (2) economics, (3) technology availability, and (4) other factors such as social factors (for example, environmental justice) and the political context. Some of those factors are discussed in key statutes under which EPA operates, and numerous EPA guidance documents on risk assessments acknowledge those factors. However, the uncertainty embedded in the data and analyses related to those factors and how those factors and their uncertainty should affect a decision are rarely discussed, other than in the statement "Once

TABLE 1-1 Continued

Statute and Program	Considerations Under the Statute
Safe Drinking Water Act (SDWA)	To regulate a contaminant under the primary drinking water regulations, EPA must first determine that the contaminant has an adverse effect on human health or welfare.[h] The maximum contaminant level goal must be established at a level at which no known health effects may occur and which allow an adequate margin of safety, without consideration of cost.[i] The mandatory, enforceable standard under the SDWA, the maximum contaminant level, is set at a level that is economically and technologically feasible. Establishing that level requires consideration of a number of factors, such as the quantifiable and nonquantifiable health risk reduction benefits,[j] the quantifiable and nonquantifiable costs,[k] and the effects of the contaminant on the general population and groups within the general population, such as infants and elderly.[l]

[a] 42 U.S.C. § 7412(b)(2)(C).
[b] 42 U.S.C. § 7412(d)(2).
[c] 42 U.S.C. § 7412(d)(5).
[d] 42 U.S.C. § 7409(b)(1).
[e] 42 U.S.C. § 7409(b)(2).
[f] 42 U.S.C. § 7410; see also Union Elec. Co. v. EPA, 427 U.S. 246, 266-67 (1976) (discussing how, although the U.S. EPA cannot consider costs in deciding whether to approve or disapprove an SIP, a state can do so in structuring its proposed SIP).
[g] 42 U.S.C. § 7503(a)(2).
[h] 42 U.S.C. § 300f(1)(B); 300f(2).
[i] 42 U.S.C. § 300g-1(b)(4)(A).
[j] 42 U.S.C. § 300g-1(b)(3)(C)(i)(I) & (II).
[k] 42 U.S.C. § 300g-1(b)(3)(C)(i)(III) & (IV).
[l] 42 U.S.C. § 300g-1(b)(3)(C)(i)(V).

the risk characterization is completed, the focus turns to communicating results to the decision maker. The decision maker uses the results of the risk characterization, other technological factors, and non-technological social and economic considerations in reaching a regulatory decision. . . . These Guidelines are not intended to give guidance on the nonscientific aspects of risk management decisions" (EPA, 2001, p. 51). In addition, other factors are referred to in Figure 1-2. Those other factors might include the political context of a decision or social factors, such as environmental justice.

Legal Context

```
                    ┌──────────────┐
                    │  Economical  │
                    │   Factors    │
                    └──────┬───────┘
                           ↓
┌──────────────┐    ┌──────────────┐    ┌──────────────┐
│ Human Health │ →  │     EPA      │ ←  │ Technological│
│ Risk Factors │    │   Decision   │    │   Factors    │
│              │    │    Making    │    │              │
└──────────────┘    └──────┬───────┘    └──────────────┘
                           ↑
                    ┌──────┴───────┐
                    │ Other Factors│
                    └──────────────┘
```

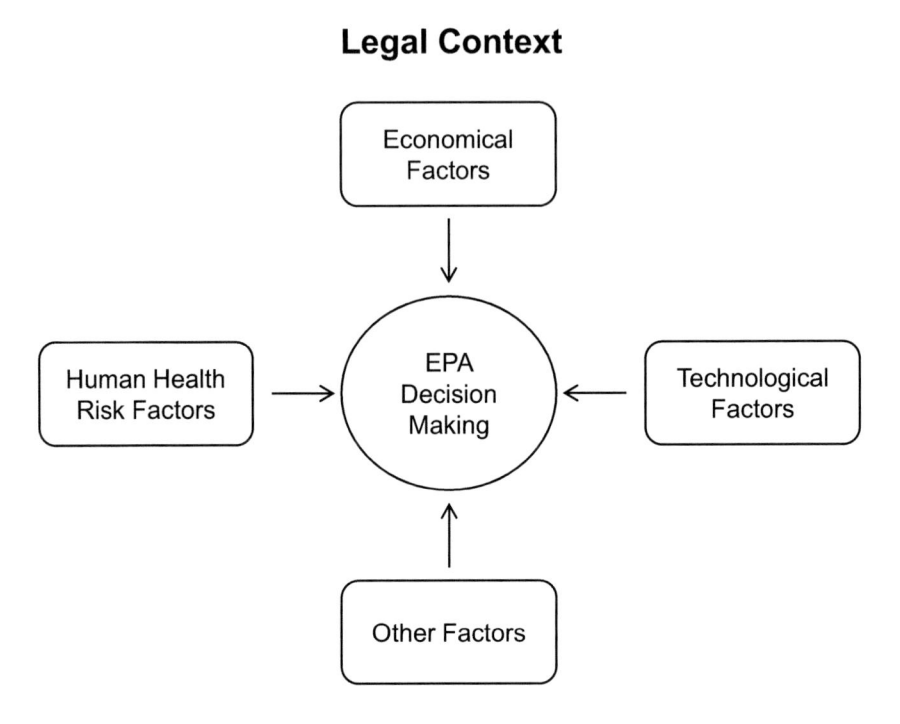

FIGURE 1-2 Factors considered in EPA's decisions.
NOTE: The legal context of a decision—that is, the statutory requirements and constraints in the nation's environmental laws—shapes the overall decision-making process, with general directives relating to data expectations, schedules and deadlines, public participation, and other considerations. That legal context also, to a large extent, determines the other factors that EPA considers in its decisions, in particular, human health risks, technological, and economical factors. At the same times those laws allow considerable discretion in the development and implementation of environmental regulations. The figure is modeled after a figure in *A Risk-Characterization Framework for Decision Making at the Food and Drug Administration* (NRC, 2011). Other factors, although more challenging to quantify, can also play a role in EPA's decisions, including social factors, such as environmental justice, and the political context.

The uncertainty in those other factors can be difficult, if not impossible, to quantify, and therefore the analysis of that uncertainty will not be discussed in depth in this report. Nevertheless, that uncertainty can influence EPA's decisions and, as discussed in Chapters 5, 6, and 7, it is important that decision makers are aware of those factors and are transparent about how those factors influenced a decision.

EPA has done a great deal of highly skilled and scientific work on uncertainty analysis for estimates of human health risks. Although the committee was not tasked with reviewing the technical aspects of those uncertainty analyses, it did review the uncertainty analyses conducted in some risk assessments for context and as examples (see Chapter 2). It also reviewed a number of guidance documents, advisory committee reports, and advice from the National Research Council (NRC), all of which, as discussed above, focus on those uncertainties dealing with risk estimates, and not on other factors that affect EPA's decisions. References to uncertainty in decision making typically discuss the uncertainty in estimates of human health risk (NRC, 1983, 1994, 1996, 2009, 2011).

The charge to the committee does not focus solely on uncertainty related to human health risk assessments but rather asks the committee to look more broadly at uncertainty in the decision-making process. For example, the charge asks how "uncertainty influence[s] risk management" and "other ways EPA can benefit from quantitative characterization of uncertainty (e.g., value of information techniques to inform research priorities)," and refers broadly to *decision making*. The committee was concerned that solely focusing attention—and resources—on reducing uncertainties in the risk assessment could lead to a false confidence that the most important uncertainties are addressed; extreme attention to reducing uncertainties in the human health risk assessment might not be sufficient without an attempt to characterize the other factors that are addressed in decision making and their uncertainties. The committee, therefore, examined the assessment of uncertainties in factors in addition to the risk to human health and the role of those uncertainties in the decision-making process (see Chapter 3). Although the uncertainty in the data and analyses related to the other factors cannot always be quantified, the report discusses the importance of being aware of those factors, potential uncertainties in those factors, and how they influence decisions, and communicating that information when discussing the rationale for EPA's decision.

When approaching its charge, the committee was aware that the consideration of uncertainty by EPA in its regulatory decisions could be evaluated in a number of ways. The uncertainty in decisions could be evaluated solely on the standard that is set—that is, whether a standard limiting the amount of ozone in air or arsenic in drinking water or establishing remediation levels for a hazardous waste site is adequate or too protective. The decisions could also be evaluated by the quality of their technical and scientific support, such as risk assessments, costs and feasibility analysis, or regulatory impact analysis. They could also be evaluated on the basis of the process by which they were developed—for example, on the opportunities for public participation, the transparency in the decision-making process, and whether social trust was established through the process. Because each of those aspects is integral to the decision-making process and each contributes to an understanding of the role of uncertainty in decision making, the committee looked broadly at those aspects when it considered decision making in the face of uncertainty.

TYPES OF UNCERTAINTY

All EPA decisions involve uncertainty, but the types of uncertainty can vary widely among decisions. For an analysis of uncertainty to be useful, a first and critical step is to identify the types of key uncertainties that are involved in a particular decision problem. Understanding the types of the uncertainty that are present will help EPA's decision makers determine when to invest resources to reduce the uncertainty and how to take that uncertainty into account in their decisions.

In this report, the committee classifies uncertainty in two categories: (1) statistical variability and heterogeneity (also called aleatory or exogenous uncertainty), and (2) model and parameter uncertainty (also called epistemic uncertainty). It also discusses a third category of uncertainty, referred to as deep uncertainty (uncertainty about the fundamental processes or assumptions underlying a risk assessment),[32] which is based on the level of uncertainty. Uncertainty stemming either from statistical variability and heterogeneity or from model and parameter uncertainty can be deep uncertainty. Chemical risk assessors typically consider uncertainty and variability to be separate and distinct, but in other fields uncertainty encompasses statistical variability and heterogeneity as well as model and parameter uncertainty (Swart et al., 2009). The committee discusses its rationale for including variability and heterogeneity as one type of uncertainty in Box 1-3.

The three different types of uncertainty are discussed below.

Statistical Variability and Heterogeneity

Variability and heterogeneity, which together are sometimes referred to as aleatory uncertainty, refer to the natural variations in the environment, exposure paths, and susceptibility of subpopulations (Swart et al., 2009). They are inherent characteristics of the system under study, cannot be controlled by decision makers (NRC, 2009; Swart et al., 2009), and cannot be reduced by collecting more information. Empirical estimates of variability and heterogeneity can, however, be better understood through research, thereby refining such estimates.

---

[32] Deep uncertainty has also been called *severe uncertainty* and *hard uncertainty* (CCSP, 2009).

> **BOX 1-3**
> **Uncertainty Versus Variability and Heterogeneity:**
> **The Committee's Use of the Terms**
>
> In the context of chemical risk assessment, uncertainty has typically been defined narrowly. For example, *Science and Decisions* (NRC, 2009) defines uncertainty as a "[l]ack or incompleteness of information" (p. 97). It defines variability as the "true differences in attributes due to heterogeneity or diversity" (p. 97), and, as with previous reports (NRC, 1983, 1994), it does not consider variability and heterogeneity to be specific types of uncertainty.
>
> In other settings, such as research on climate change, variability is considered a type or nature of uncertainty (CCSP, 2009). In addition, even reports related to chemical risk assessment highlight the importance of evaluating both uncertainty and variability in risk assessments and considering both in the decision-making process (NRC, 1983, 1994, 2009). When EPA makes a regulatory decision it must consider what information it might be missing, the variability and heterogeneity in the information that it has, and the uncertainty in that variability and heterogeneity. The committee, therefore, discusses variability and heterogeneity in this report. When using the term *uncertainty* generically, it includes variability and heterogeneity in its definition.

Variability occurs within a probability distribution that is known or can be ascertained. It can often be quantified with standard statistical techniques, although it may be necessary to collect additional data. If variability comes, in part, from heterogeneity, populations can be divided into subcategories on the basis of demographic, economic, or geographic characteristics with associated percentages and possibly a probability distribution over the percentages if they are uncertain. That stratification into distinct categories can tighten the probability distribution within each of the categories. Variability in the underlying parameters often depends on personal characteristics, geographic location, or other factors, and there might not be an adequate sample size to detect true underlying differences in populations or to ensure that data are sufficiently representative of the population being studied.

There are many variables outside of the decision maker's control that can affect the appropriateness of a particular decision or its consequences (for example, socioeconomic factors or comorbidities). Modeling those factors is not always feasible. For example, there may be too many socioeconomic factors that require large samples to analyze, insufficient time to conduct an appropriate statistical survey and analysis, or a prohibition of using some sociodemographic variables in the analysis. A longer-term research agenda can often evaluate such variables, and retrospective evaluations of specific decisions can help identify their effects and improve the decision-making process.

Model and Parameter Uncertainty

Model[33] and parameter uncertainty, which together constitute epistemic uncertainty, include uncertainty from the limited scientific knowledge about the nature of the models that link the causes and effects of environmental risks with risk-reduction actions as well as uncertainty about the specific parameters of the models. There may be various disagreements about the model, such as which model is most appropriate for the application at hand, which variables should be included in the model, the model's functional form (that is, whether the relationship being modeled is linear, exponential, or some other form), and how generalizable the findings based on data collected in another context are to the problem at hand (for example, the generalizability of findings based on experiments on animals to human populations). Such disagreements add to the uncertainty when making a decision on the basis of that information, and, therefore, the committee considers scientific disagreements when discussing uncertainty.

In theory, model and parameter uncertainty can be reduced by additional research (Swart et al., 2009). Issues related to model specification can sometimes be resolved by literature reviews and by various technical approaches, including meta-analysis. It is often necessary to extrapolate well beyond the observations used to derive a given set of parameter estimates, and the functional form of the model used for that extrapolation can have a large effect on model outputs. Linear or curvilinear forms may yield substantially different projections, and theory rarely provides guidance about which functional form to choose. The best approaches involve reexamining the fit of various functional forms within the data that are observed, comparing projections based on alternative functional forms, or both. Those approaches, however, are not always fruitful.

Using data from a population that is similar to the population to which the policy will be applied helps address generalizability. In practice this may involve using observational data on human populations to supplement data from controlled experiments in animals. Both animal and human data have important advantages and disadvantages, and ideally both would be

---

[33] A *model* is defined as a "simplification of reality that is constructed to gain insights into select attributes of a particular physical, biologic, economic, or social system. Mathematical models express the simplification in quantitative terms" (NRC, 2009, p. 96). Model parameters are "[t]erms in a model that determine the specific model form. For computational models, these terms are fixed during a model run or simulation, and they define the model output. They can be changed in different runs as a method of conducting sensitivity analysis or to achieve a calibration goal" (NRC, 2009, p. 97).

available. In some cases, the data may yield similar answers or it may be possible to interpret the differences. Experts in a given field or domain can sometimes quantify model and parameter uncertainty either by drawing on existing research, by relying on their professional experience, or a combination of the two. Such an approach is not a substitute for research, but if done, it should build on the best data, models, and estimates available, using experts to integrate and quantify this knowledge. EPA's peer-review process and its consultation with the Science Advisory Board are examples of this. Those approaches are discussed in Chapter 5.

Deep Uncertainty

Deep uncertainty is uncertainty that it is not likely to be reduced by additional research within the time frame that a decision is needed (CCSP, 2009). Deep uncertainty typically is present when underlying environmental processes are not understood, when there is fundamental disagreement among scientists about the nature of the environmental processes, and when methods are not available to characterize the processes (such as the measurement and evaluation of chemical mixtures). When deep uncertainty is present, it is unclear how those disagreements can be resolved. In other words, "Deep uncertainty exists when analysts do not know, or the parties to a decision cannot agree on, (1) the appropriate models to describe the interactions among a system's variables, (2) the probability distributions to represent uncertainty about key variables and parameters in these models and/or (3) how to value the desirability of alternative outcomes" (Lempert et al., 2003, p. 3).

In situations characterized by deep uncertainty, the probabilities associated with various regulatory options and associated utilities may not be known. Deep uncertainty often occurs in situations in which the time horizon for the decision is unusually long or when there is no prior record that is relevant for analyzing a problem following a major unanticipated event, such as climate change or the migration of radionuclides from a geological repository over the coming 100,000 years. There can also be substantial disagreement about base-case (that is, the case with no intervention) exposure levels; about the rate of economic growth; or about future changes in the relationship between economic activity and exposures, losses associated with an adverse outcome, or how to value such losses—that is, the utility levels given to different magnitudes of losses. For example, experts could disagree about the effects that an oil spill in a body of water has on human health, quality of life, animal populations, and employment. Box 1-4 presents an example of a decision made in the face of deep uncertainty.

Neither the collection nor the analysis of data nor expert elicitation to assess uncertainty is likely to be productive when key parties to a decision

> **BOX 1-4**
> **An Example of a Decision in the Face of Deep Uncertainty**[a]
>
> Decisions that had to be made in the United Kingdom during the outbreak of bovine spongiform encephalopathy (BSE) in British cattle in the mid-1980s illustrate deep uncertainty in a decision. Public health officials had to decide whether potential risks from the fetal calf serum in the media in which vaccines were prepared warranted halting childhood immunization programs against infectious diseases (such as diphtheria and whooping cough). In describing the context of the decision, Sir Michael Rawlins stated:
>
>> There was very little science to go on. The presumption that scrapie and BSE were caused by the same prion was just that, a presumption. Experiments to confirm or refute this would take two or three years to complete. There was no information as to whether maternal–fetal transmission of the prion took place. There was no evidence of whether or not vaccines prepared using fetal calf serum contained infected material. Again it would take two or three years to find out. There was no idea of a dose–response relationship and this probability of development of disease. And again two years would pass before such information would be available.
>
> A quick decision was needed despite the deep uncertainty surrounding the risks of creating an epidemic of BSE in infants; there was no time to decrease that uncertainty. The unknown risks of BSE had to be weighed against the risks of "the re-appearance of diphtheria, whooping cough, and other infectious diseases in children." The risk of lethal infectious diseases by abandoning the immunization program for three years "approached 100 percent." The decision was made to continue the immunization program; luckily, there was no outbreak of diseases related to BSE. In his speech, Sir Rawlins points out that they reached the correct conclusions, albeit perhaps not for the right reasons. But the moral of the tale is this, all decisions in the fields in which we work require good underpinning science but they also always require judgments to be made.
>
> ---
>
> [a] Sir Michael Rawlins described this decision in his acceptance speech for the International Society for Pharmacoeconomics and Outcomes Research Avedis Donebedian Outcomes Research Lifetime Achievement Award on May 21, 2011, in Baltimore, Maryland. The text of the speech is available at http://www.ispor.org/news/articles/July-Aug2011/avedis.asp (accessed March 17, 2012).

do not agree on the system model, prior probabilities, or the cost function. The task is to make decisions despite the presence of deep uncertainty using the available science and judgment, to communicate how those decisions were made, and to revisit those decisions when more information is available.

OVERVIEW OF THE REPORT

As discussed above, uncertainty analysis at EPA has focused on the uncertainty in estimates of risks to human health. In Chapter 2 the committee briefly discusses how EPA evaluates and considers uncertainty in those risk estimates, using case studies to illustrate the effects of those uncertainties on EPA's decisions. The committee then moves beyond that narrow focus of uncertainty in EPA's decisions and in Chapter 3 looks at the uncertainty inherent in other factors that play a role in EPA's regulatory decisions, that is, the economic, technological, social, and political factors. In Chapter 4 the committee examines decision making in other public health policy settings to examine whether the tools and techniques used in those fields could improve EPA's decision-making processes. In Chapter 5 the committee applies the information discussed in the previous three chapters to the context of EPA's regulatory decisions, discussing how uncertainty in the different factors that affect EPA's decisions should be incorporated into the agency's decision-making process, incorporating uncertainty into the three-phase framework presented in *Science and Decisions: Advancing Risk Assessment* (NRC, 2009). The committee recommends approaches to evaluating the various uncertainties and to prioritizing which uncertainties to account for in EPA's different decisions. Details of some of those approaches are presented in Appendix A. Scientific expertise is necessary for uncertainty analysis to guide decision making, but it is not enough. Transparency, consideration of community values, and inclusion of stakeholders are prerequisites to ultimately building trust in the decisions that are reached and leading to acceptance of governmental decisions (Kasperson et al., 1999; NRC, 1989, 1996; Presidential/Congressional Commission on Risk Assessment and Risk Management, 1997a). Chapter 6 focuses on those aspects of EPA's regulatory decision-making process. Chapter 7 discusses the practical implications of this report and includes the recommendations that derive from the discussions in Chapter 2 through Chapter 6. Appendixes B and C contain biographical sketches of committee members and the agendas of the committee's open sessions, respectively.

REFERENCES

Budescu, D. V., S. Broomell, and H. H. Por. 2009. Improving communication of uncertainty in the reports of the Intergovernmental Panel on Climate Change. *Psychological Science* 20(3):299–308.

CCSP (Climate Change Science Program). 2009. *Best practice approaches for characterizing, communicating, and incorporating scientific uncertainty in decisionmaking*. [M. Granger Morgan (lead author), Hadi Dowlatabadi, Max Henrion, David Keith, Robert Lempert, Sandra McBride, Mitchell Small, and Thomas Wilbanks (contributing authors)]. A report by the Climate Change Science Program and the Subcommittee on Global Change Research. National Oceanic and Atmospheric Administration.

EPA (U.S. Environmental Protection Agency). 1993. Re: Quantitative uncertainty analysis for radiological assessment. http://yosemite.EPA.gov/sab/sabproduct.nsf/8FE9A83C1BE1BA7A85257323005C134F/$File/ANALYSIS%20ASSESS%20%20RAC-COM-93-006_93006_5-9-1995_86.pdf (accessed November 20, 2012).
———. 1999a. An SAB report on the National Center for Environmental Assessment's comparative risk framework methodology. http://yosemite.EPA.gov/sab/sabproduct.nsf/83F6D5FD42385D46852571930054E70E/$File/dwc9916.pdf (accessed November 20, 2012).
———. 1999b. An SAB report: Estimating uncertainties in radiogenic cancer risk. http://yosemite.EPA.gov/sab/sabproduct.nsf/D3511CC996FB97098525718F0064DD44/$File/rac9908.pdf (accessed November 20, 2012).
———. 2000. *Risk characterization handbook*. Washington, DC: Science Policy Council, EPA.
———. 2001. Guidelines for developmental toxicity risk assessment. *Federal Register* 56(234):63798–63826.
———. 2007. Benefits and costs of Clean Air Act—direct costs and uncertainty analysis. http://yosemite.EPA.gov/sab/sabproduct.nsf/598C5B0B5A89799A852572F4005C406E/$File/council-07-002.pdf (accessed November 20, 2012).
———. 2010a. Guidelines for preparing economic analyses. http://yosemite.epa.gov/ee/epa/eed.nsf/pages/guidelines.html (accessed January 10, 2013).
———. 2010b. Review of EPA's microbial risk assessment protocol. http://yosemite.EPA.gov/sab/sabproduct.nsf/07322F6BB8E5E80085257746007DC64F/$File/EPA-SAB-10-008-unsigned.pdf (accessed November 20, 2012).
———. 2011. Our mission and what we do. U.S. Environmental Protection Agency. http://www.EPA.gov/aboutepa/whatwedo.html (accessed November 20, 2012).
Kasperson, R. E., D. Golding, and J. X. Kasperson. 1999. Trust, risk, and democratic theory. In *Social trust and the management of risk*, edited by G. Cvetkovich and R. Lofstedt. London: Earthscan. Pp. 22–44.
Lempert, R. J., S. W. Popper, and S. C. Bankes. 2003. *Shaping the next one hundred years: New methods for quantitative, long-term policy analysis*. RAND Corporation.
Lindley, D. 2006. *Uncertainty: Einstein, Heisenberg, Bohr, and the struggle for the soul of science*. New York: Random House.
NRC (National Research Council). 1983. *Risk assessment in the federal government: Managing the process*. Washington, DC: National Academy Press.
———. 1989. *Improving risk communication*. Washington, DC: National Academy Press.
———. 1994. *Science and judgment in risk assessment*. Washington, DC: National Academy Press.
———. 1996. *Understanding risk: Informing decisions in a democratic society*. Washington, DC: National Academy Press.
———. 2002. *Estimating the public health benefits of proposed air pollution regulations*. Washington, DC: The National Academies Press.
———. 2009. *Science and decisions: Advancing risk assessment*. Washington, DC: The National Academies Press.
———. 2011. *A risk-characterization framework for decision making at the Food and Drug Administration*. Washington, DC: The National Academies Press.
OMB (Office of Management and Budget). 2011. *Report to Congress on the benefits and costs of federal regulations and unfunded mandates on state, local, and tribal entities*. Washington, DC: OMB.
Presidential/Congressional Commission on Risk Assessment and Risk Management. 1997a. *Risk assessment and risk management in regulatory decision-making. Final report.* Volume 1. Washington, DC: Government Printing Office.
———. 1997b. *Risk assessment and risk management in regulatory decision-making. Final report. Volume 2.* Washington, DC: Government Printing Office.

Swart, R., L. Bernstein, M. Ha–Duong, and A. Petersen. 2009. Agreeing to disagree: Uncertainty management in assessing climate change, impacts and responses by the IPCC. *Climatic Change* 92:1–29.

Wallsten, T. S., and D. V. Budescu. 1995. A review of human linguistic probability processing: General principles and empirical evidence. *The Knowledge Engineering Review* 10(01):43–62.

Wallsten, T. S., D. V. Budescu, A. Rapoport, R. Zwick, and B. Forsyth. 1986. Measuring the vague meanings of probability terms. *Journal of Experimental Psychology: General* 155(4):348–365.

2

Risk Assessment and Uncertainty

As discussed in Chapter 1, a number of factors play a role in the decisions made by the U.S. Environmental Protection Agency's (EPA's) decisions. This chapter discusses the uncertainty in the data and the analyses associated with one of those factors, human health risk estimates. There has been a great deal of progress over the past few decades in developing methods to assess and quantify uncertainty in estimates of exposure, adverse effects, and overall risks (EPA, 2004). In this chapter the committee provides a broad overview of the nature of the main uncertainties in the characterization of risks. Later chapters offer discussions of how EPA should incorporate those uncertainties into its decisions and communicate them. This chapter begins with background information on risk assessments and then summarizes various approaches to characterizing the uncertainties in risk estimates. Examples of EPA's risk assessments and the uncertainty analyses in them are then discussed.

RISK ASSESSMENT

The mandate of EPA is broad. It includes regulating the releases and human exposures arising at any stage of manufacture, distribution, use, and disposal of any substances that pose environmental risks. In the context of the EPA's mandate, the various risks to health arise because of the presence of chemicals and other agents, such as radiation-emitting substances and pathogenic microorganisms, in different media, including air, water, and soils. The chemicals of interest include industrial products of diverse types and by-products of chemical manufacturing, chemical use, and energy

> **BOX 2-1**
> **Definitions**
>
> *Human health risk assessment* is a systematic framework within which scientific information relating to the nature and magnitude of threats to human health is organized and evaluated. The typical goal of a human health risk assessment is to develop a statement regarding the likelihood, or probability, that exposures arising from a given source, or in some cases from multiple sources, will harm human health (NRC, 1983). The risks to a given population are a function of the hazards of a given chemical and the exposure that the population experiences.
>
> *Risk communication* "is an interactive process of exchange of information and opinion among individuals, groups, and institutions. It involves multiple messages about the nature of risk and other messages, not strictly about risk, that express concerns, opinions, or reactions to risk messages or to legal and institutional arrangements for risk management" (NRC, 1989, p. 21).
>
> *Risk management* refers to the process whereby the results of a risk assessment are considered, together with the results of other technical analyses and nonscientific factors, to reach a decision about the need for and extent of risk reduction to be sought in particular circumstances and of the means for achieving and maintaining that reduction (NRC, 1983). At the EPA, risk management is typically linked to a regulatory decision, whereas risk assessment involves the evaluation of the scientific evidence about risks that inform that regulatory decision. As discussed in *Science and Decisions: Advancing Risk Assessment* (NRC, 2009), a conceptual distinction between risk assessment and risk management is maintained as it is "imperative that risk assessments used to evaluate risk-management options not be inappropriately influenced by the preferences of [decision makers]" (p. 12).

production. The EPA uses a health risk assessment and risk-management model to identify the nature and estimate the magnitude of risks from chemicals and other agents and to determine the best way to manage or mitigate those risks (EPA, 2004). As discussed in Chapter 1, the process of risk assessment and using it for regulatory decisions was first described in the seminal 1983 National Academy of Sciences report *Risk Assessment in The Federal Government: Managing the Process* (NRC, 1983) (hereafter the Red Book)[1] and in a series of expert reports issued since that time (NRC, 1994, 1996, 2007, 2009). All of those reports emphasize the need for a conceptual distinction between risk assessment and risk management. Box 2-1 offers descriptions of some of the important terms in this area.

---

[1] The National Research Council study that led to the Red Book was congressionally mandated and was requested "to strengthen the reliability and objectivity of scientific assessment that forms the basis for federal regulatory policies applicable to carcinogens and other public health hazards" (NRC, 1983, p. iii).

The scientific information about the hazards used in risk assessments is derived largely from observational epidemiology and experimental animal studies of specific substances or combinations of substances that are designed to identify their hazardous properties (that is, the types of harm they can induce in humans) and the conditions of exposure under which those harms are observed (that is, the dose and duration) (Box 2-2 provides

---

**BOX 2-2
Development of Estimates of Human Health
Risks for Non-Cancer Endpoints**[a]

When assessing the risks to human health from a chemical for a non-cancer endpoint, EPA typically develops a reference dose (RfD).[b] EPA defines an RfD as an "estimate (with uncertainty spanning perhaps an order of magnitude) of a daily oral exposure to the human population (including sensitive subgroups) that is likely to be without an appreciable risk of deleterious effects during a lifetime" (EPA, 2012b). The RfD is based on the assumption that a certain dose must be exceeded before toxicity is expressed. The RfD is derived from a no-observed-adverse-effect level (NOAEL), a lowest-observed-adverse-effect level (LOAEL), or a benchmark dose from animal or epidemiology studies. The NOAEL is the "highest exposure level at which there are no biologically significant increases in the frequency or severity of adverse effect between the exposed population and its appropriate control" (EPA, 2012b). The LOAEL is the "lowest exposure level at which there are biologically significant increases in frequency or severity of adverse effects between the exposed population and its appropriate control group" (EPA, 2012b). The benchmark dose is a "dose or concentration that produces a predetermined change in response rate of an adverse effect (called the benchmark response or BMR) compared to background" (EPA, 2012b). In general, NOAELs and LOAELs are derived from animal data; benchmark doses are derived from epidemiologic studies.

In developing the RfD, the NOAEL, LOAEL, or benchmark dose is generally reduced downward by uncertainty factors (UFs) which are usually multiples of 10, to account for limitations and incompleteness in the data. Those limitations could include knowledge of interspecies variability, and the expectation that variability in response in the general population is likely to be much greater than that present in the populations (human or animal) from which the NOAEL, LOAEL, or benchmark dose is derived. Whether standard defaults or data-based uncertainty factors are used, the accuracies of the UFs used are largely unknown, so quantitative characterization of the uncertainties associated with any given RfD is generally not possible.

---

[a] The processes for non-cancer and cancer risk assessments are not static. A number of reports, including *Science and Decision: Advancing Risk Assessment* (NRC, 2009), have recommended harmonizing the processes used for non-cancer and cancer risk assessments.
[b] A reference concentration (RfC) is developed for inhalation toxicants.

> **BOX 2-3**
> **Development of Estimates of Human Health**
> **Risks for Cancer Endpoints**[a]
>
> In March 2005, EPA updated its guidelines for estimating the human health risks associated with a carcinogen (EPA, 2005a).[b] Those guidelines are briefly summarized here. The first step in a cancer risk assessment is to characterize the hazard using a "weight of evidence narrative."[c] The narrative describes the available evidence, including its strengths and limitations, and "provides a conclusion with regard to human carcinogenic potential" p. 1-12). The data available for each tumor type are then used to derive a point of departure (POD), that is, "an estimated dose (usually expressed in human-equivalent terms) near the lower end of the observed range, without significant extrapolation to lower doses" (EPA, 2005a, p. 1-13). Data from epidemiology studies are used if available and of sufficient quality. In the absence of such epidemiology data, data from animal studies are used and, when possible and appropriate, toxicokinetic data are used to inform cross-species dose scaling to estimate the human-equivalent dose. The POD is generally "the lower 95% confidence limit on the lowest dose level that can be supported for modeling by the data" (EPA, 2005a).
>
> Once the POD is established, extrapolation is used to model the dose–response relationship at exposures lower than the POD. Depending on how much is known about the mode of action[d] of the agent, one of two methods is used for the extrapolation: linear or nonlinear extrapolation.
>
> A linear extrapolation is used in the "absence of sufficient information on modes of action" or when "the mode of action information indicates that the dose-response curve at low dose is or is expected to be linear" (EPA, 2005a, p. 1-15). For a linear extrapolation, "a line should be drawn from the POD to the origin, corrected for background" (EPA, 2005a, p. 3-23). The slope of that line, called the slope factor, is considered "an upper-bound estimate of risk per increment of dose" (EPA, 2005a, p. 3-23) and is used to estimate risks at different exposure levels.

a description of how those data are used for non-cancer endpoints, and Box 2-3 gives a description for cancer endpoints). Information from these studies is used to develop the hazard identification and dose–response (where "response" is the harm or adverse effect) components of a risk assessment. The data used to develop these components typically arise from diverse sources and types of study designs and frequently lack strong consistency in methods so that reaching valid conclusions about them requires both careful scientific evaluations and experienced judgments. A hallmark of the modern risk-assessment framework is the expectation not only that the scientific evidence is described, but also that the evaluation of the evidence and any judgments about the quality and relevance of the evidence to the risk assessors are thoroughly and clearly described (OMB and OSTP, 2007).

A nonlinear approach is used "when there are sufficient data to ascertain the mode of action and conclude that it is not linear at low doses and the agent does not demonstrate mutagenic or other activity consistent with linearity at low doses." "[F]or nonlinear extrapolation the POD is used in the calculation of a *reference dose* [RfD]or *reference concentration* [RfC]" (EPA, 2005a, p. 3-16) similar to how an RfD or RfC is estimated for non-cancer endpoints, as described in Box 2-2.

Depending on the amount of information available about potential susceptible populations and susceptibility during different life stages, adjustments to the estimates or separate assessments are recommended in the guidelines. Concurrent with the release of the general cancer risk assessment guidelines, EPA released supplemental guidelines that provide "specific guidance on procedures for adjusting cancer potency estimates only for carcinogens acting through a mutagenic mode of action" (EPA, 2005a, p. 1-19).

[a] The processes for non-cancer and cancer risk assessments are not static. A number of reports, including *Science and Decision: Advancing Risk Assessment* (NRC, 2009), have recommended harmonizing the processes used for non-cancer and cancer risk assessments.
[b] The "cancer guidelines are not intended to provide the primary source of, or guidance for, the Agency's evaluation of the carcinogenic risks of radiation" (EPA, 2005a, p. 1-6).
[c] EPA recommends using one of the five standard hazard descriptors: Carcinogenic to Humans, Likely to Be Carcinogenic to Humans, Suggestive Evidence of Carcinogenic Potential, Inadequate Information to Assess Carcinogenic Potential, and Not Likely to Be Carcinogenic to Humans (EPA, 2005a).
[d] Mode of action "is defined as a sequence of key events and processes, starting with interaction of an agent with a cell, proceeding through operational and anatomical changes, and resulting in cancer formation. A "key event" is an empirically observable precursor step that is itself a necessary element of the mode of action or is a biologically based marker for such an element. Mode of action is contrasted with "mechanism of action," which implies a more detailed understanding and description of events, often at the molecular level, than is meant by mode of action" (EPA, 2005a, p. 1-10).

Assessing exposure requires an evaluation of the nature of the population that is incurring exposures to the substances of interest and the conditions of exposure that it is experiencing (such as the dose and duration of exposure) (NRC, 1991). In effect, risk to the exposed population is understood by examining the exposure the population experiences (its "dose") relative to the hazard and dose–response information described above. Risk characterization consists of a statement regarding the "response" (risk of harm) expected in the population under its exposure conditions, together with a description of uncertainties (NRC, 1983). Risk assessments are frequently used by EPA to characterize health risks under existing exposure conditions and also to examine how risks will change if actions are taken to alter exposures (EPA, 2012b). A clear description of the confidence that can be placed in the risk-assessment result—that is, a statement regarding the scientific uncertainties associated with the assessment—should be a feature of all risk assessments.

UNCERTAINTY AND RISK ASSESSMENT

Uncertainties are inherent in all scientific undertakings and cannot be avoided. The extent to which uncertainties in data and analyses can be measured and expressed in highly quantitative terms depends upon the types of investigations used to develop scientific knowledge. Highly controlled experiments, usually conducted in a laboratory or clinical setting, if well designed and conducted, can provide the clearest information regarding uncertainties. Even in many experimental studies, however, it is not always possible to quantify uncertainties. Controlled clinical trials, for example, still contain uncertainties and variability that cannot necessarily be predicted or accurately quantified. Using available knowledge with its inherent uncertainties to make predictions about as-yet unobserved—and perhaps inherently unobservable—states is even more uncertain, but it is critical to many important social decisions, including EPA's decisions related to human health protection (EPA, 2012b). Risk assessments can address such questions as whether a risk to health will be reduced if certain actions are taken and, if so, by what magnitude and whether new risks might be introduced when such actions are taken. However, the scientific uncertainties associated with such predictive efforts include not only the uncertainty associated with the available knowledge but also uncertainty related to the predictive nature of estimates (for example, predicting how much of a decrease in air pollution different control technologies will produce or predicting how many lung cancer cases will be avoided by a given decrease in air pollution). The Red Book highlighted many of the unknowns in a risk assessment, including a lack of understanding of the mechanisms that underlie different adverse effects (NRC, 1983). The presence of uncertainty in data and analyses, however, is not unique to the chemical risk-assessment world and should not preclude a regulatory decision. For instance, drugs are often used even without a thorough understanding of their underlying mechanism of action.

Understanding Risk: Informing Decisions in a Democratic Society (NRC, 1996) emphasizes the importance to decision making of recognizing uncertainties in risk assessments, pointing out that decision makers should attempt to consider "both the magnitude of uncertainty and its sources and character" (p. 5). The report further emphasizes, however, that "unrecognized sources of uncertainty—surprise and fundamental ignorance about the basic processes that drive risk—are often important sources of uncertainty" (NRC, 1996, p. 5). Because of that, the report argues that the limitations in uncertainty analyses should be recognized and considered and that the focus of any such analysis should be on the uncertainties that most affect the decision, and it criticizes characterizations of risks that do not focus on the questions of greatest impact to the decision outcome.

Uncertainties in data and analyses can enter the risk-assessment process at every step; the sources of the largest uncertainties include the use of observational studies, extrapolation from studies in animals to humans, extrapolation from high- to low-dose exposures, and interindividual variability. Box 2-4, which briefly describes the evidence on the degreasing solvent trichloroethylene (TCE), provides an example of how uncertainties arise in risk assessments and of the challenges that those uncertainties present to decision makers.

Studies in humans that evaluate whether exposure to a substance causes specific adverse effects can provide the most relevant information on hazards and dose response. Clinical trials have a greater chance of yielding unambiguous results regarding causality than do observational studies (Gray-Donald and Kramer, 1988). It is not ethical to intentionally expose people to chemicals at exposure concentrations that are likely to cause adverse effects, even following a short duration of exposure. Moreover, clinical trials are costly and typically are designed to capture the short-term effects of an intervention, whereas many adverse effects of chemicals can take decades to develop. Except under highly limited conditions, clinical trials should not be used to study the adverse health effects of substances regulated by EPA (NRC, 2004). Most studies evaluating risks in humans, therefore, are observational in nature; that is, they investigate some aspect of the physical world "as it is."

Observational studies can have significant limitations. Because many such studies do not provide evidence that meets the criteria typically used to establish causation rather than association—that is, the Hill criteria, such as demonstrating a dose response and a temporal relationship between exposure and effect (Hill, 1965)—the results from individual observational studies on their own can, at best, be used to establish associations. For example, in many situations the only information is whether or not participants were exposed to a given chemical, and nothing is known about the magnitude of individual exposures or whether there was differential exposure among individuals, which makes it very difficult to determine dose–response relationships. Observational studies often capture exposures and health outcomes retrospectively, so that the temporal relationship between the exposure and the outcome cannot be determined. Furthermore, regardless of the study type, inconsistent results in a group or body of studies examining a given chemical are common and contribute to uncertainties regarding causality. The types of uncertainties associated with the interpretation of results from observational studies may be described quantitatively—such as conducting

> **BOX 2-4**
> **Trichloroethylene Risk Assessment: An Example of the Uncertainties Present in a Cancer Risk Assessment and How They Could Affect Regulatory Decisions**
>
> Trichloroethylene (TCE) is a degreasing solvent used in many industries and a contaminant in all environmental media (air, water, soil). The issues in the TCE risk assessment illustrate several uncertainties and related choices that risk assessors and decision makers face when evaluating the risk potential of environmental carcinogens, as well as the delays that such uncertainties can lead to. They also highlight the resulting reliance on assumptions and models in the absence of definitive data, the need for choices among the options that exist due to unknowns and uncertainties, and the role of these uncertainties and choices in shaping regulatory decisions. The evidence related to TCE, described briefly below, has been summarized previously (EPA, 2009; NRC, 2005). Sources of uncertainty in the assessment include the following:
>
> - Although human studies provide evidence of associations between occupational exposures to TCE and liver and kidney cancers, and non-Hodgkin's lymphoma (see EPA, 2009, Chapter 4), there is uncertainty about whether those associations are causal.
> - If the associations are assumed causal, there is uncertainty in the cancer potency (that is, the risk of cancer per unit of exposure to TCE). Estimates of cancer potency based on data from different human studies differ by up to 100-fold (EPA, 2009; NRC, 2005).

an analysis that estimates the effect and also quantitative assessments of the likelihood of that effect—but the uncertainties are usually expressed in qualitative language, such as describing the range of relative risks across studies and the quality of the individual studies.

Data from experimental studies in animals and from a variety of in vitro systems are commonly used, in part, to overcome the limitations in observational epidemiology studies. Experimental studies allow researchers to acquire information about hazards and dose response and, if well designed and well performed, can yield information about causality. Results from such studies, however, can have significant uncertainties regarding their generalizability to humans. Differences between the metabolism and the mode of action of a chemical in animals compared with humans underlie many of the differences between animals and humans, but uncertainty often exists about the magnitude of the differences. For example, it is not currently possible to quantify the extent to which disease processes observed in animal experiments apply to humans, and differences in longevity have to

- Data from animal studies indicate that TCE can induce liver and lung cancer (mice) and kidney and testicular cancers (rats). Estimates of cancer potencies derived from the animal data differ by over 500-fold (EPA, 2009).
- Potency differences based on animal data are explained in part by the use of different models for low-dose extrapolation, but the current understanding of the biological mechanisms of cancer induction is too limited to allow a selection of the optimal model (EPA, 2009; NRC, 2005).
- The biological reasons for the differences in response between animals and humans are only partially understood, resulting in uncertainty about which studies (animal or human), and which potency estimates (at the lower or higher end of the range) are more reliable and about the nature and extent of possible human risk in populations exposed through the environment (EPA, 2009).

The choices risk assessors make when interpreting the data in light of the uncertainty influence the size of the risk estimate and, in turn, the decision whether or not to regulate TCE and, if so, the nature of the regulatory standards that are based on the risk assessment. For example, if assessors use potency values at the lower end of the range, the assessment may indicate a low likelihood of cancer risk in humans and obviate the need for regulatory action. By contrast, if assessors use potency values at the higher end of the range, the assessment may indicate a high likelihood of cancer risk in humans and be the basis for a more stringent regulatory standard.

be taken into account when considering the duration of exposure for animal studies. It is important to note, however, that despite those limitations enough is known about the similarities and differences between humans and experimental animals to make them relevant to and critical for assessing human health risks (EPA, 2011a).

There is also uncertainty associated with exposure information. One such uncertainty comes from extrapolating from exposures in studies to the exposures experienced by the public. There are instances in which the exposure incurred by the population that is the subject of a risk assessment (that is, the target population) is close to, or even in the same range as, that for which hazard and dose–response data have been collected. For example, studies of exposures to the primary air pollutants ozone, lead, mononitrogen oxides, sulfur oxide gases, and particulate matter are often in the same ranges of exposures as occurs with the general population (Dockery et al., 1993; Pope et al., 1995). In many instances, however, the exposure incurred by the target population is only a small fraction—sometimes a very tiny fraction—of the exposures for which it has been possible to collect hazard and dose–response information. Studies of occupational cohorts, for example, typically involve exposures well in excess of general population exposures, and animal studies similarly involve high-dose effects. For the risks to the target population to be described, a method or model must be used to extrapolate from the high-dose scientific findings to infer the risks at much lower doses. That extrapolation can create large uncertainties in risk assessment. The biological bases for selecting among different models for extrapolation are not well established, and different models can yield different estimates of low-dose risk.

In other cases very little might be known about the actual exposures in the target population, adding additional uncertainty. Individuals within a population also vary with respect to both their exposures and their responses to hazardous substances. Reliable, quantitative information that allows an understanding of the magnitudes of that variability can be difficult, if not impossible, to acquire (Samoli et al., 2005). Risk assessments need to account for possible differences in response between the populations that were studied to understand hazards and dose response in the target population, which typically is more diverse than the population studied (Pope, 2000). Studies of human exposure in limited populations cannot be used to apply to other, more diverse populations without considering the uncertainties from the different populations.

Those uncertainties are part of almost all risk assessments conducted by EPA (EPA, 2004). Additional uncertainties related to the effects of chemicals at different life stages and different comorbidities, the effects of exposures to complex mixtures, and the effects of chemicals that have received very little toxicological study are also introduced in many assessments (EPA, 2004). In many cases the analyst or scientist who is conducting the risk assessment is able only to describe those uncertainties in largely qualitative terms, and formulating scientifically rigorous statements about the effects of these uncertainties on a risk result is beset with difficulties (EPA, 2004).

THE HISTORY OF UNCERTAINTY ANALYSIS

The 1983 NAS Report, Uncertainties and the Use of Defaults

Given the EPA's mandate to protect human health, the agency has had to find a way to make decisions taking into account the scientific uncertainty discussed above. The Red Book emphasized that the uncertainties inherent in risk assessment were so pervasive that virtually no risk assessment could be completed without the use of assumptions or some types of models for extrapolations (NRC, 1983). Moreover, it recognized that there was little or no scientific basis for discriminating among the range of assumptions or models that might be used in a given case. Given that situation, risk assessments were not likely to achieve any degree of consistency and, indeed, might be easily "tailored" to meet any risk-management objective. The report argued that some degree of general scientific understanding, though limited, exists in each of the areas of uncertainty that attend risk assessment. It further argued that, in many of the areas of uncertainty, a range of plausible scientific inferences might be made, although none could be claimed to be generally correct (that is, correct for all or most specific cases). If the agencies conducting risk assessments could select, for each step where one was needed, the "best supported" option or inference and could apply that inference to all of its risk assessments, then it could be possible to be consistent in risk assessment and to minimize case-by-case manipulations. Determining the "best" option cannot be based upon science alone, but also requires a policy choice, and agencies needed to specify clearly the scientific and policy bases for their choices among available options. The report further stated that the selected set of inference options for risk assessment should not only be justified, but also be set down in written guidelines for the conduct of risk assessments, so that they could be visible to all (NRC, 1983).

As recommended, EPA has developed guidelines for the conduct of risk assessments for many types of adverse effects, and those guidelines include recommendations about what uncertainty factors to use when there are specific uncertainties (EPA, 1986, 1992, 1997a,b,d, 1998a, 2004, 2005a). The selected sets of inference options have come to be called *uncertainty factors*, or *defaults*. In practice, in reviewing the scientific information available on specific substances or exposures, it becomes clear that there are significant gaps in knowledge or information; agency human health risk assessors adopt the relevant default specified in the guidelines. For example, to account for uncertainties in how to extrapolate from animal data to risks in humans, the default uncertainty factor is 10. EPA, therefore, divides the dose at which no effect is seen in animals by a factor of 10 to estimate a dose at which an effect would not be seen in humans. If there are data on the extent of toxicokinetic differences between animals and humans, then EPA might use a data-derived uncertainty factor rather than using the default uncertainty factor.

The Problems with Default-Driven Risk Assessments

In addition to helping make risk assessments consistent across agencies, the use of prespecified, generic defaults has a number of advantages. First, although the uncertainties and limitations in the estimate should be characterized for the decision maker, the use of a default does allow the assessor to provide a risk estimate when a decision needs to be made in the presence of some uncertainty. The assessor can use a standard default to extrapolate when there is little or no scientific information available to indicate what the shape of the dose–response curve is in the low-dose region for a carcinogen; the default in this case would be a *linear, no-threshold model*. Second, defaults are typically protective of health. EPA originally selected the linear, no-threshold default as a "conservative" or "health-protective" policy choice because it assumes that there is no dose below which risks are not increased. It is likely to generate the highest, or upper-bound, risk estimate consistent with the data; the actual risk almost certainly will not exceed the upper bound and will likely fall below it. Third, it can provide decision makers with a single, upper-bound point estimate, while acknowledging the uncertainty in that point estimate by indicating that the actual risk could fall anywhere between zero and that upper bound. If that upper bound is itself in the negligible risk range, the uncertainty statement allows the decision maker to assert that any actual risks are likely to be below the negligible range. Fourth, the use of a single point estimate and defaults allows for a simpler risk-communication message.

Using defaults to deal with uncertainty does, however, have a number of deficiencies, and that use has been the subject of much discussion and debate in the scientific literature (NRC, 1994, 2006, 2009). Defaults have been criticized for their lack of an adequate scientific basis. For example, the National Research Council (NRC) criticized EPA's use of defaults in its dioxin risk assessment, in one instance stating that EPA's use of the "default linear model lacked adequate scientific support" (NRC, 2006). In addition, if the fact that they are used and the implications of their use are not communicated with a risk estimate, they can mask the uncertainty, providing a sense of uncertainty that is inaccurate. The use of defaults has also been criticized for being overly conservative; that is, the regulatory standards that are based on defaults are more restrictive than necessary to protect public health. If, as with the linear, no-threshold default, most of the defaults in risk assessments are selected because they are conservative (that is, protective of health and resulting in lower permissible exposures or emissions), very little can be said about exactly how much uncertainty is associated with their cumulative use. Indeed, even in the case of the example of the upper-bound estimate yielded by the use of the linear, no-threshold default, there remain significant questions about how much greater the upper bound is than the true risk and whether the differences are at all consistent across different risk assessments—that is, whether there is any basis for believing that the upper-bound estimate for one substance has the same relation to the "true" risk as it does for another substance. As discussed below, these and other criticisms have led to suggestions for alternative ways to treat the problems of uncertainty in risk assessment.

The Use of Data, Not Defaults

The Red Book recognized the limitations of defaults and also recognized that any set of defaults, no matter how they were selected, would not likely be generally applicable to all risk assessments (NRC, 1983). Although substantial research might someday make it possible to justify generally applicable models for interspecies, high- to low-dose, or intraspecies extrapolations, the understanding needed to achieve such a goal remains unavailable and is not likely to be available for a very long time.

New research on a specific substance or exposure situation can lead to questions about the applicability of any given default to that substance or situation (NRC, 1983). Thus the report urged agencies conducting risk assessments to seek data that would supplant the need for a default—such as data on the toxicokinetic differences between animals and humans—and to allow scientific knowledge and data on specific substances to hold sway over defaults. This same point has been emphasized in subsequent reports (NRC, 1994, 2007, 2009; OMB and OSTP, 2007).

For instance, if enough scientific information exists about the differences in the metabolism or mode of action of a chemical in animals versus in humans, then scientifically derived extrapolation factors can be used rather than the defaults. Such factors, which EPA refers to as "data-derived extrapolation factors" (EPA, 2011a, p. ii), would be specific for a given chemical. If those factors more accurately reflect the differences between animals and humans than default adjustment factors, the use of such data-derived extrapolation factors would decrease the uncertainty in the risk assessment.

EPA agrees with the NRC report that specific knowledge should supplant the use of defaults when appropriate and it has adopted that as a general principle (EPA, 2005a). However, a 2006 GAO report concluded that "EPA is often reluctant to deviate from its established default assumptions" (GAO, 2006, p. 67). In other words, GAO concluded that although EPA in theory favors using new scientific information to supplant established defaults, in practice it uses defaults more often than not (GAO, 2006).

The continued reliance on defaults is, in part, due to a view that any research data used to deviate from defaults—such as data on the mode of action of a chemical that indicates that there are no adverse effects below a certain dose—will themselves have uncertainties. Unless those uncertainties are clearly much smaller than those associated with the default, assessors often think that the default should be retained (Haber et al., 2001; Meek et al., 2002). However, because the true uncertainties associated with the standard defaults are generally unknowable, such comparisons are problematic. In any event, the general question remains unanswered of just how convincing the data on specific substances—such as mode of action data—should be in order to use the specific information rather than a default. The debate about whether default adjustment factors or specific data should be used occurs even for individual chemicals or other agents. This has occurred, for example, with the dioxin risk assessment (NRC, 2006) and the formaldehyde risk assessments (NRC, 2011).

Several National Academy of Sciences (NAS) committees have recommended that EPA develop explicit criteria for when to use research data on specific substances rather than defaults to deal with uncertainties and have also recommended that EPA minimize use of defaults (NRC, 1994, 2009). For example, *Science and Decisions: Advancing Risk Assessments* (hereafter *Science and Decisions*; NRC, 2009) stated that "criteria should be available for judging whether, in specific cases, data are adequate for direct use or to support an inference in place of a default" (NRC, 2009, p. 7).

NEWER APPROACHES TO DEALING WITH UNCERTAINTIES

A number of NAS reports (NRC, 1994, 1996) and much research provide guidance and methods for uncertainty analyses that can replace the use of default adjustment factors. And, indeed, EPA has adopted a number of those methods to evaluate the uncertainty in many of the components (data and analyses) of its human health risk assessments.

For exposure assessments, much progress has been made in moving away from "point estimates" and toward characterizing population exposures in terms of distributions, and EPA now routinely uses such approaches for some areas of exposure assessment (EPA, 1997b,d). Significant developments have also occurred in characterizing the magnitude of uncertainty in certain types of hazard and dose–response information, particularly for the primary air pollutants (NRC, 2002).

As discussed in *Science and Decisions* (NRC, 2009), EPA uses a number of different approaches to quantify the uncertainty and variability in different components of a risk assessment. One such method is Monte Carlo analysis, a technique that propagates uncertainty—including variability and heterogeneity as well as model and parameter uncertainty—in the various components of the human health risk assessment (for example, in the exposures, toxicokinetics, and the dose response). The techniques can incorporate a range of values and propagate that range throughout the assessment to create a distribution of risk estimates rather than a single point risk estimate. (See the discussion of risk assessment for arsenic in drinking water below for an example.) Such efforts are leading to more complete—but more complex—characterizations of uncertainty than have traditionally been provided to decision makers. The techniques can be combined with Bayesian techniques, in which one assesses not simply the potential range of values but also the likelihood of a given value within the range based on expert judgment and available information (NRC, 2009). This technique has been used, for example, to estimate mortality data from air pollution (Zeger et al., 2004).

Although such analyses can provide valuable information, as detailed in the 1996 report *Understanding Risk: Informing Decisions in a Democratic Society* (NRC, 1996) and other reports (for example, NRC, 2009), the risk assessment and its uncertainty analyses should be decision-driven activities. The emphasis on the use of risk assessment to inform choices and to evaluate different decision options plays a major role in the 2009 NRC report. As the report explains, a critical aspect of the effort to make risk assessment more useful is to ensure that the level of analytical effort devoted to specific risk assessments is appropriate to the decision context (NRC, 2009). It follows that the level and extent of uncertainty analysis needed for specific risk assessments is also dependent upon the decision context. Considering and planning for the uncertainty analysis during the initial problem formulation stage is essential to making sure that the analysis is done in the context of the decision.

To increase the likelihood of achieving sound and acceptable decisions, *Understanding Risk* (NRC, 1996) recommended that decision-making organizations implement an "analytical–deliberative" process in which the characterization of risk emerges from a combination of analysis and deliberation among decision makers, analytic experts (such as natural and social scientists), and interested and affected parties (such as legislators, environmentalists, industry groups, citizens' groups, and others) (NRC, 1996). This is different from former practices that often involved risk assessors conducting their assessments without any input from stakeholders. Subsequent NRC reports (NRC, 1989, 1999, 2005) have also suggested processes that integrate science with public participation to enhance the quality of environmental assessments and decisions. Later chapters in this report also emphasize the need to integrate public participation into the decision-making process, including the risk-assessment stage.

EXAMPLES OF EPA'S RISK ASSESSMENTS

During its deliberations the committee referred to and reviewed a number of EPA's risk assessments to determine how it conducts and uses uncertainty analyses. Below, the committee briefly summarizes three risk assessments—of arsenic in drinking water, the Clean Air Interstate Rule (CAIR), and of methylmercury—to highlight how EPA incorporates uncertainty analyses into its estimates of health risks and uses that information in its decisions.

Arsenic in Drinking Water

The regulation of arsenic in drinking water illustrates the quantitative approach EPA has used in estimating health risks.

In 1976 the EPA proposed an interim maximum contaminant level (MCL) for arsenic of 50 micrograms per liter (µg/L). In 1996, as part of a review of that MCL, EPA requested that "the National Research Council (NRC) independently review the arsenic toxicity data base and evaluate the scientific validity of EPA's 1988 risk assessment for arsenic in drinking water" (NRC, 1999, p. 1). The resulting 1999 report, *Arsenic in Drinking Water* (NRC, 1999), concludes that "the current EPA MCL for arsenic in drinking water of 50 µg/L does not achieve EPA's goal for public-health protection and, therefore, requires downward revision as promptly as possible" (p. 9). It further recommended sensitivity analyses to examine the uncertainty from the choice of dose–response model, and to evaluate the uncertainty from measurement error, confounding, and nutritional factors.

On January 22, 2001, EPA issued a final rule for arsenic in drinking water, with a pending standard for arsenic of 10 µg/L (EPA, 2001b). That standard was developed by relying on the scientific information in the 1999 NRC report, and was set based on risks of bladder and lung cancer. The agency estimated risks by using a linear extrapolation of data from an epidemiological study of exposures in southwestern Taiwan. To explore the uncertainty created by using different models of the dose–response relationship, EPA compared estimates calculated by Morales et al. (2000) using 10 different models and chose the model that did not result in a supralinear extrapolation because there was no biological basis for such an extrapolation. There was also uncertainty about exposures caused by variability in how much water people drink, including differences between the U.S. population and the Taiwanese population in the study and differences within the U.S. population. A Monte Carlo analysis was used to estimate distributions of water intake, accounting for age, sex, and weight and adjusting water intake to account for the high consumption of water from cooking in Taiwan, and the mortality data in Taiwan were converted to expected incident data in the United States. EPA also stated that because of the increased intake of water on a per-body-weight basis in infants fed formula, it intended to issue a health advisory for the use of low-arsenic water in the preparation of infant formula.

On April 23, 2001, under a new administration, the agency announced that it would delay the effective date of the arsenic and drinking water rule and that it had asked the NAS to review the data, including any new data since the publication of *Arsenic in Drinking Water* (NRC, 1999) on the health effects of arsenic exposure (EPA, 2001c). The agency also asked the National Drinking Water Advisory Council to review the cost estimates for the rule and its Science Advisory Council to review the arsenic rule benefits analysis. (See Chapter 3 for further discussion of cost and benefit analyses.)

The NAS report, *Arsenic in Drinking Water: Update 2001* (NRC, 2001) confirmed that EPA's human health risk assessment should focus on bladder and lung cancer and that it should be based on the epidemiologic data from southwestern Taiwan. The report recommended using an "additive Poisson model with a linear term used for dose" (p. 215) to extrapolate from the doses in the epidemiology study to the lower exposures seen in the United States. Based on a determination that the available information on mode of action did not indicate an appropriate method for extrapolating, it recommended that a default linear extrapolation should be used. It noted, however, that the choice to use a linear extrapolation is, in part, a policy decision (NRC, 2001).

The report also discussed the effects of other uncertainties and evaluated the effect of using different studies (for example, one with data on populations in Chile); statistical models, including using a model-weighting approach; background incidence rates between different populations; and water intakes and measurement error. The report presented maximum-likelihood estimates (that is, central tendencies), not upper-bound or worst-case estimates. The report (NRC, 2001) concluded "that recent studies and analyses enhance the confidence in risk estimates" (p. 14), and that the results of the updated assessment "are consistent with the results presented in the NRC's 1999 *Arsenic in Drinking Water* report and suggest that the risks for bladder and lung cancer incidence are greater than the risk estimates on which EPA based its January 2001 pending rule" (p. 14). It also discussed the uncertainty that could come from variability in arsenic metabolism, different exposures, nutritional parameters, and interactions between arsenic and smoking that could affect the dose–response curve.

On October 31, 2001, EPA announced that it would set the arsenic in drinking water standard at 10 µg/L and not delay the implementation schedule first established in the January 22, 2001, regulation (EPA, 2001a).

The example of arsenic in drinking water illustrates the broad spectrum of uncertainty and sensitivity analyses that can be conducted when estimating human health risks. The effects of those evaluations provide a broader view of how uncertainty about background rates of cancer, water intake, model choice, and data from studies can affect the risk estimates. They also, however, indicate how those uncertainties do not always affect the estimates to an extent that would affect the overall decision. For example, despite the uncertainties listed and the different health risk estimates presented and additional work by a second NRC committee, those new data and analyses also supported the original 10 µg/L standard promulgated in January 2001. The example also illustrates the importance that political factors can play in EPA's decisions. Despite the characterization and quantification of uncertainty in the 1999 report (NRC, 1999) on which EPA's January 2001 rule was based, a new administration called into question the scientific basis of the rule and required a reevaluation of the science.

Clean Air Interstate Rule

In 2005, EPA published its regulatory impact analysis (RIA) for CAIR, a rule developed to implement requirements of the Clean Air Act concerning the transport of air pollution across state boundaries (EPA, 2005b). A December, 2008 court ruling directed EPA to issue a new that rule, but did not vacate CAIR.[2] In response to that ruling, in July 2011, EPA issued the Cross-State Air Pollution Rule (CSAPR) to implement the cross-state pollution transportation requirements of the CAA. In August 2012, the U.S. Court of Appeals for the D.C. Circuit ruled that the CSAPR violates Federal law and must be vacated because: (1) "EPA has used the good neighbor provision to impose massive emissions reduction requirements on upwind States without regard to the limits imposed by the statutory text" (p. 7); and (2) "when EPA quantified States' good neighbor obligations, it did not allow the States the initial opportunity to implement the required reductions with respect to sources within their borders. Instead, EPA quantified States' good neighbor obligations and *simultaneously* set forth EPA-designed Federal Implementation Plans, or FIPs, to implement those obligations at the State level. By doing so, EPA departed from its consistent prior approach to implementing the good neighbor provision and violated the Act" (p. 7).[3] EPA is reviewing that court decision and CAIR remains in place (EPA, 2012a). The committee discusses the uncertainty analyses contained in the RIA below.

In 2005, EPA published its RIA for CAIR in which it presented the benefits and the costs of the rule, and the comparative costs of implementing CAIR in 2010 and 2015 (EPA, 2005b). As discussed by Krupnick et al. (2006), EPA conducted a number of uncertainty and sensitivity analyses in support of that rulemaking. They used two different approaches to characterizing uncertainties in health benefits: one based on "the classical statistical error expressed in the underlying health effects and economic valuation studies used in the benefits modeling framework" (p. 1-6) and one using the results of a "pilot expert elicitation project designed to characterize key aspects of uncertainty in the ambient $PM_{2.5}$/mortality relationship,

---

[2] State of North Carolina v. Environmental Protection Agency, 05-1244. U.S. App. D.C. (2008) (http://www.EPA.gov/airmarkets/progsregs/cair/docs/CAIRRemandOrder.pdf [accessed June 8, 2012]).

[3] EME Homer City Generation L.P. v. Environmental Protection Agency, et al., 11-1302. U.S. App. D.C. (2012) (http://www.cadc.uscourts.gov/internet/opinions.nsf/19346B280C784 05C85257A61004DC0E5/$file/11-1302-1390314.pdf [accessed June 8, 2012]).

and augments the uncertainties in the mortality estimate with the statistical error reported for other endpoints in the benefit analysis" (EPA, 2005b, p. 1-6). EPA also used two different social discount rates (3 percent and 7 percent) to estimate the social benefits and costs of the rule. They point out a number of uncertainties that were not captured in the analyses, including model specification, emissions, air quality, the likelihood that particulate matter causes premature mortality, and other health effects. In reviewing the analysis and presentation of uncertainty in the RIA, Krupnick et al. (2006) pointed out that three pages of the executive summary of the RIA are devoted to discussing uncertainties, but criticized the report because "the summary tables do not include ranges for estimates of benefits or indicate that the reported numbers represent a mean of a distribution, nor does the section reporting out health benefits include any mention of uncertainty" (p. 58). They also point out that, as in many RIAs, EPA qualitatively discusses "uncertainties in each section but leaving any quantitative information in the appendices" (Krupnick et al., 2006, p. 58).

The uncertainty analyses, however, focus to a large extent on the uncertainties in the health benefits, and not the uncertainties in costs and technological factors. As EPA (2005) states, the cost estimates

> assume that all States in the CAIR region fully participate in the cap and trade programs that reduce SO2 and NOx emissions from EGUs. The cost projections also do not take into account the potential for advancements in the capabilities of pollution control technologies for SO2 and NOx removal and other compliance strategies, such as fuel switching or the reductions in their costs over time. EPA projections also do not take into account demand response (i.e., consumer reaction to electricity prices) because the consumer response is likely to be relatively small, but the effect on lowering private compliance costs may be substantial. Costs may be understated since an optimization model was employed and the regulated community may not react in the same manner to comply with the rules. The Agency also did not factor in the costs and/or savings for the government to operate the CAIR program as opposed to other air pollution compliance programs and transactional costs and savings from CAIR's effects on the labor supply. (p. 1-5)

Methylmercury

Mercury (Hg) is converted to methylmercury by aquatic biota, and it bioaccumulates in aquatic food webs. Methylmercury can lead to neurotoxic effects in humans, and consumption of large, predatory fish is the major source of human exposure to methylmercury. Under the CAA Amendments of 1990,[4] EPA had to determine whether it is "appropriate

---

[4] CAA Amendments of 1990, Pub. L. No. 101-549 Sec. 112(n)(1)(A) (1990).

and necessary" to regulate the release of "air toxics" from electric-utility steam-generating units[5] (hereafter, a power plant) prior to regulating the release of Hg from those plants. In 1997 EPA published a *Mercury Study Report to Congress* (EPA, 1997c), and in 1998 it published a *Study of Hazardous Air Pollutant Emissions from Electric Utility Steam Generating Units* (EPA, 1998b). The former examined "mercury emissions by source, the health and environmental implications of those emissions, and the availability and cost of control technologies" (EPA, 1997c, p. O-1). The latter includes "(1) a description of the industry; (2) an analysis of emissions data; (3) an assessment of hazards and risks due to inhalation exposures to 67 hazardous air pollutants (HAPs); (4) assessments of risks due to multipathway (inhalation plus non-inhalation) exposures to four HAPs (radionuclides, mercury, arsenic, and dioxins); and (5) a discussion of alternative control strategies" (EPA, 1998b, p. ES-2). However, because of gaps in the scientific data regarding Hg toxicity, Congress directed EPA[6] to have the NAS conduct a study on the health effects of Hg. Specifically, NAS was to evaluate EPA's RfD estimating the health effects of methylmercury.

When NAS began its study, EPA, the U.S. Food and Drug Administration (FDA), and the Agency for Toxic Substances and Disease Registry (ATSDR) all had published risk assessments that used different methods and relied on different studies for their estimates of health risks. The estimates of a "safe" level of exposure from the three different agencies were an RfD of 0.1 microgram/kg/day from EPA, an action level of 0.5 microgram/kg/day from FDA, and a minimal risk level of 0.3 microgram/kg/day from ATSDR.

In its evaluation NAS focused on three epidemiologic studies and evaluated their strengths and weaknesses in detail (NRC, 2000). Two studies—one conducted in the Faroe Islands (Grandjean et al., 1997) and one conducted in New Zealand (Kjellström et al., 1986, 1989)—concluded that there was an association between in utero exposure to methylmercury from maternal fish consumption and an increased risk of poor scores on neurobehavioral test batteries in early childhood. A third study—conducted in the Seychelles (Davidson et al., 1998)—concluded that no such association existed.

NAS identified and analyzed a number of uncertainties in the scientific evidence, including the following (NRC, 2000):

---

[5] An electric-utility steam-generating unit was defined as "any fossil-fuel–fired combustion unit of more than 25 megawatts electric (MWe) that serves a generator that produces electricity for sale."
[6] Departments of Veterans Affairs and Housing and Urban Development, and Independent Agencies Appropriations Act of 1999, Pub. L. No. 105-276 (1999).

- *Uncertainty related to benchmark doses.* To compare benchmark doses generated by the three different studies, the committee analyzed the data for multiple endpoints from each of the three studies using the same statistical techniques and presented the range of benchmark doses generated from the different analyses. To compare the effect on the benchmark dose of using a single study versus analyzing data from all three studies together, the committee estimated and presented a benchmark dose by conducting an integrative analysis using Bayesian statistical approaches.
- *Uncertainty related to default factors.* To determine whether to use a default uncertainty-adjustment factor (a factor of 10 is the default) to account for variability among humans, the committee reviewed the toxicokinetic data on methylmercury measurements. After examining the scientific evidence the committee recommended against using the default uncertainty-adjustment factor for toxicokinetic variability and recommended a factor of two to three.
- *Uncertainty related to human variability among subpopulations.* After looking at potentially sensitive populations, the committee highlighted the need to consider susceptible populations, including pregnant women and subsistence fishermen, in the assessment and subsequent decisions (NRC, 2000).

Since the publication of that report, EPA has conducted a regulatory impact analysis and published a rule to regulate the release of mercury and other toxic substances from coal-fired power plants (EPA, 2011b). As detailed in EPA's regulatory impact analysis in support of the final standards, the agency used a Bayesian hierarchical statistical model that integrates the data from the epidemiology studies for its dose–response model. That model, which was published by Axlerad et al. (2007), draws on the integrative analysis conducted by NAS (NRC, 2000). The analyses also show the effect of inclusion and exclusion of a potential outlier in one of the studies. The analyses in the regulatory impact analyses also included risk and benefit calculations related to concomitant decreases in particles less than 2.5 micrometers in diameter ($PM_{2.5}$) emissions from the emission-control technologies that would be put in place. Unlike the NAS report (NRC, 2000), however, EPA does not present the effects of various choices on the estimates of health risks. Given that in the regulatory impact analysis much of the monetized benefits come from co-benefits due to decreased $PM_{2.5}$-related premature mortalities, the lack of detailed uncertainty analyses for mercury might be appropriate. The estimated benefits from $PM_{2.5}$ reductions are presented as a range ($37 billion to $990 billion + B), with the lower and higher benefits calculated using mortality estimates from two different published studies, and B representing an amount from benefits that were not quantified.

Although few detailed, quantitative uncertainty analyses are presented for the risk estimates for mercury exposures, the regulatory impact analysis does note a number of uncertainties in the analysis. Those uncertainties include "selection of IQ as a primary endpoint when there may be other more sensitive endpoints, selection of the blood-to-hair ratio for mercury, [sic] the dose–response estimates from the epidemiological literature [, and c]ontrol for confounding from the potentially positive cognitive effects of fish consumption and, more specifically, omega-3 fatty acids" (EPA, 2011b, pp. E-17–E-18).

The regulatory impact analysis also discusses, and in some cases analyzes, uncertainties in factors other than health risk estimates that contribute to EPA's decisions, such as economic, technological, and social factors (EPA, 2011b); those are discussed in Chapter 3.

KEY FINDINGS

- Uncertainty in the data and analyses that are used in the assessment of risks is inescapable. Decision makers need to understand—either quantitatively or qualitatively—the types and magnitude of the uncertainty that are present in order to make an informed decision.
- Consideration of uncertainty analyses for the human health risk assessment should begin during the initial stages of considering a decision to help ensure that the analyses are appropriate to the decision.
- Although the use of agent-specific research-based adjustments is preferable, it is sometimes necessary and acceptable to use default adjustment factors to account for uncertainty in human health risk assessments. For example, defaults might need to be used when research-based analysis could lead to prolonged delays in regulatory decisions.
- Regardless of whether agent-specific research-based factors or default adjustment factors are used, communicating the basis of adjustment factors and their impact on human health risk estimates to decision makers and stakeholders is critical for regulatory decisions.
- EPA has made great strides in assessing the uncertainties in risk estimates, for example, by developing and applying probabilistic techniques and Monte Carlo analysis to uncertainty analysis.
- Although some uncertainty analyses are required by statute, the analyses conducted are not always helpful in agency decisions, and

in some cases, such as dioxin, striving to analyze every uncertainty might delay regulatory decisions.
- Consideration of uncertainty analysis should include the perspectives of stakeholders, and should be useful to the decision makers.

RECOMMENDATION 1
To better inform the public and decision makers, U.S. Environmental Protection Agency (EPA) decision documents[7] and other communications to the public should systematically

- include information on what uncertainties in the health risk assessment are present and which need to be addressed,
- discuss how the uncertainties affect the decision at hand, and
- include an explicit statement that uncertainty is inherent in science, including the science that informs EPA decisions.

REFERENCES

Axelrad, D. A., D. C. Bellinger, L. M. Ryan, and T. J. Woodruff. 2007. Dose–response relationship of prenatal mercury exposure and IQ: An integrative analysis of epidemiologic data. *Environmental Health Perspectives* 115(4):609–615.
Davidson, P. W., G. J. Myers, C. Cox, C. Axtell, C. Shamlaye, J. Sloane-Reeves, E. Cernichiari, L. Needham, A. Choi, and Y. Wang. 1998. Effects of prenatal and postnatal methylmercury exposure from fish consumption on neurodevelopment. *Journal of the American Medical Association* 280(8):701–707.
Dockery, D. W., C. A. Pope, X. Xu, J. D. Spengler, J. H. Ware, M. E. Fay, B. G. Ferris Jr., and F. E. Speizer. 1993. An association between air pollution and mortality in six US cities. *New England Journal of Medicine* 329(24):1753–1759.
EPA (Environmental Protection Agency). 1986. *Guidelines for carcinogen risk assessment.* Washington, DC: Risk Assessment Forum, EPA.
———. 1992. *Guidelines for exposure assessment.* Washington, DC: Risk Assessment Forum, EPA.
———. 1997a. *Guidance on cumulative risk assessment: Part 1 planning and scoping.* http://www.EPA.gov/brownfields/html-doc/cumrisk2.htm (accessed January 14, 2012).
———. 1997b. *Guiding principles for Monte Carlo analysis.* Washington, DC: Risk Assessment Forum, EPA.
———. 1997c. *Mercury study report to Congress.* Washington, DC: EPA.
———. 1997d. *Policy for use of probabilisitc analysis in risk assessment.* http://www.EPA.gov/osa/spc/pdfs/probpol.pdf (accessed April 15, 2012).
———. 1998a. *Guidelines for neurotoxicity risk assessment.* Washington, DC: Risk Assessment Forum, EPA.
———. 1998b. *Study of hazardous air pollutant emissions from electric utility steam generating units—final report to Congress.*

---

[7] The committee uses the term "decision document" to refer to EPA documents that go from EPA staff to the decision maker and documents produced to announce an agency decision.

———. 2001a. EPA announces arsenic standard for drinking water of 10 parts per billion, edited by EPA. Washington, DC: EPA.

———. 2001b. National primary drinking water regulations; arsenic and clarifications to compliance and new source contaminants monitoring. *Federal Register* 66(14):6976–7066.

———. 2001c. National primary drinking water regulations; arsenic and clarifications to compliance and new source contaminants monitoring. *Federal Register* 66(78):20580–20584.

———. 2004. *An examination of EPA risk assessment principles and practices.* Washington, DC: Office of the Science Advisor, EPA.

———. 2005a. *Guidelines for carcinogen risk assessment.* Washington, DC: Risk Assessment Forum, EPA.

———. 2005b. *Regulatory impact analysis for the final Clean Air Interstate Rule.* Washington, DC: Office of Air and Radiation, EPA.

———. 2009. *Toxicological review of Trichloroethylene: In support of summary information on the Integrated Risk Information System (IRIS) [External Review Draft].* Washington, DC: EPA.

———. 2011a. *Draft—guidance for applying quantitative data to develop data-derived extrapolation factors for interspecies and intraspecies extrapolation.*

———. 2011b. *Regulatory impact analysis for the final mercury and air toxics standards.* Research Triangle, NC: Environmental Protection Agency, Office of Air Quality Planning and Standards, Health and Environmental Impacts Division.

———. 2012a. *Clean Air Interstate Rule (CAIR).* http://www.EPA.gov/cair/index.html (accessed November 11, 2012).

———. 2012b. *EPA risk assessment—basic information.* http://EPA.gov/riskassessment/basicinformation.htm#arisk (accessed June 12, 2012).

GAO (Government Accountability Office). 2006. *Human health risk assessment: EPA has taken steps to strengthen its process, but improvements needed in planning, data development, and training.* Washington, DC: GAO.

Grandjean, P., P. Weihe, R. F. White, F. Debes, S. Araki, K. Yokoyama, K. Murata, N. Sørensen, R. Dahl, and P. J. Jørgensen. 1997. Cognitive deficit in 7-year-old children with prenatal exposure to methylmercury. *Neurotoxicology and Teratology* 19(6):417–428.

Gray-Donald, K., and M. Kramer. 1988. Causality inference in observational vs. experimental studies. *American Journal of Epidemiology* 127(5):885–892.

Haber, L. T., J. S. Dollarhide, A. Maier, and M. L. Dourson. 2001. Noncancer risk assessment: Principles and practice in environmental and occupational settings. In *Patty's toxicology*: John Wiley & Sons, Inc.

Hill, A. B. 1965. The environment and disease: Association or causation? *Proceedings of the Royal Society of Medicine* 58:295–300.

Kjellström, T., P. Kennedy, S. Wallis, and C. Mantell. 1986. *Physical and mental development of children with prenatal exposure to mercury from fish. Stage 1: Preliminary tests at age 4.* Vol. Report 3080. Solna, Sweden: National Swedish Environmental Protection Board.

Kjellström, T., P. Kennedy, S. Wallis, and C. Mantell. 1989. *Physical and mental development of children with prenatal exposure to mercury from fish. Stage II: Interviews and psychological tests at age 6.* Solna, Sweden: National Swedish Environmental Protection Board.

Krupnick, A., R. Morgenstern, M. Batz, P. Nelson, D. Burtraw, J. Shih, and M. McWilliams. 2006. *Not a sure thing: Making regulatory choices under uncertainty.* Washington, DC: Resources for the Future.

Meek, M., A. Renwick, E. Ohanian, M. Dourson, B. Lake, B. Naumann, and V. Vu. 2002. Guidelines for application of chemical-specific adjustment factors in dose/concentration–response assessment. *Toxicology* 181:115–120.

Morales, K. H., L. Ryan, T.-L. Kuo, M.-M. Wu, and C.-J. Chen. 2000. Risk of internal cancers from arsenic in drinking water. *Environmental Health Perspectives* 108(7):655–661.

NRC (National Research Council). 1983. *Risk assessment in the federal government: Managing the process.* Washington, DC: National Academy Press.
———. 1989. *Improving risk communication.* Washington, DC: National Academy Press.
———. 1991. *Human exposure assessment for airborne pollutants: Advances and opportunities.* Washington, DC: National Academy Press.
———. 1994. *Science and judgment in risk assessment.* Washington, DC: National Academy Press.
———. 1996. *Understanding risk: Informing decisions in a democratic society.* Washington, DC: National Academy Press.
———. 1999. *Arsenic in drinking water.* Washington, DC: National Academy Press.
———. 2000. *Toxicological effects of methylmercury.* Washington, DC: National Academy Press.
———. 2001. *Arsenic in drinking water: 2001 update.* Washington, DC: National Academy Press.
———. 2002. *Estimating the public health benefits of proposed air pollution regulations.* Washington, DC: The National Academies Press.
———. 2004. *Intentional human dosing studies for EPA regulatory purposes: Scientific and ethical issues.* Washington, DC: The National Academies Press.
———. 2005. *Risk and decisions about disposition of transuranic and high-level radioactive waste.* Washington, DC: The National Academies Press.
———. 2006. *Health risks from dioxin and related compounds.* Washington, DC: The National Academies Press.
———. 2007. *Scientific review of the proposed risk assessment bulletin from the Office of Management and Budget.* Washington, DC: The National Academies Press.
———. 2009. *Science and decisions: Advancing risk assessment.* Washington, DC: The National Academies Press.
———. 2011. *Review of the Environmental Protection Agency's draft IRIS assessment of formaldehyde.* Washington, DC: The National Academies Press.
OMB (Office of Management and Budget) and OSTP (Office of Science and Technology Policy). 2007. *Updated principles for risk analysis: Memorandum for the heads of executive departments and agencies.* http://www.whitehouse.gov/omb/memoranda/fy2007/m07-24.pdf (accessed January 4, 2012).
Pope, C. A., M. J. Thun, M. M. Namboodiri, D. W. Dockery, J. S. Evans, F. E. Speizer, and C. W. Heath, Jr. 1995. Particulate air pollution as a predictor of mortality in a prospective study of us adults. *American Journal of Respiratory and Critical Care Medicine* 151(3):669–674.
Pope III, C. A. 2000. Review: Epidemiological basis for particulate air pollution health standards. *Aerosol Science & Technology* 32(1):4–14.
Samoli, E., A. Analitis, G. Touloumi, J. Schwartz, H. R. Anderson, J. Sunyer, L. Bisanti, D. Zmirou, J. M. Vonk, and J. Pekkanen. 2005. Estimating the exposure–response relationships between particulate matter and mortality within the APHEA multicity project. *Environmental Health Perspectives* 113(1):88.
Zeger, S., F. Dominici, A. McDermott, and J. M. Samet. 2004. Bayesian hierarchical modeling of public health surveillance data: A case study of air pollution and mortality. *Monitoring the health of populations: Statistical principles and methods for public health surveillance.* Oxford: Oxford University Press. Pp. 267–288.

3

Uncertainty in Technological and Economic Factors in EPA's Decision Making

In Chapter 1 the committee specified three factors that affect decisions made by the U.S. Environmental Protection Agency (EPA): estimates of human health risks, technology availability, and economics (see Figure 1-2). As outlined in Chapter 1, the legal context within which a decision is made determines, to a large extent, which of those factors are considered by EPA. EPA's analyses of uncertainty have traditionally focused on the uncertainties in human health risk estimates (discussed in Chapter 2), but uncertainties in technological and economic factors can affect EPA's decisions. In this chapter the committee discusses the uncertainty in those factors. Although these three factors are not independent—for example, the estimates of human health risks influence the economic analysis, and technology availability contributes to economic analyses—the committee discusses them separately.

In addition to those three factors, other factors, including the political climate and social factors such as environmental justice and public sentiment, can also affect EPA's decisions. Although there is uncertainty about such factors, that uncertainty is difficult, if not impossible, to quantify. Those factors, therefore, are discussed separately at the end of this chapter.

TECHNOLOGY AVAILABILITY

Technology Assessments

Congress recognizes that technological considerations—including the feasibility, impacts, and range of risk-management options—are key to many of EPA's regulatory issues and decisions, and many of the statutes that grant EPA its regulatory authority require the evaluation of the technology available to implement a proposed regulation. Exactly how Congress directs EPA to consider technological factors differs from statute to statute and, in some cases, from section to section in a single statute (see Box 3-1 for a description of different technology standards). Statutes typically require a consideration of the technologies likely to be feasible within a given time frame, and some require engineering costs (that is, the costs of purchasing, installing, and operating the technologies to meet the standard) to be considered when assessing feasibility.[1] EPA typically considers—and court decisions have supported that consideration[2]—"technologies that are currently available" to include technologies that can reasonably be anticipated to be developed in the future.

The section of the Clean Air Act (CAA) related to mobile emission sources requires EPA to evaluate the "availability of technology, including costs"[3] when considering revised tailpipe emission standards (Section 202), but to evaluate the "best technology that can reasonably be anticipated to be available at the time such measures are to be implemented"[4] when considering standards for urban buses (Section 219). When EPA is deciding whether to list a pollutant as a hazardous air pollutant (HAP) under the CAA, it is only allowed to consider "adverse effects on human health or adverse environmental effects,"[5] not cost or technical feasibility. When it sets standards for HAPs under the CAA, however, in addition to health risk assessments, it is required to take the maximum achievable control technology into account. The CAA also requires EPA to establish National Ambient Air Quality Standards solely on the basis of health risks, and it is up to individual states to set implementation plans to achieve those standards.

Under the Clean Water Act (CWA), EPA is required to develop effluent standards for sources of water pollution based on the "best practicable control technology currently available"[6] (BPT), or the "best available technology economically achievable"[7] (BAT). The CWA establishes the framework for determining the BAT for conventional water pollution.[8] In general, BAT effluent limits represent the best available economically achievable

---

[1] The initial estimation of engineering costs could be considered part of the technology assessment or part of the benefit–cost assessment. Regardless, those costs will be included in the economic analysis in the regulatory impact assessment.
[2] See, for example, Portland Cement Assn. v. Ruckelshaus, 486 F.2s 375 (D.C. Circ. 1973).
[3] 42 U.S.C. § 7521 (i)(2)(A)(i) (2012).
[4] 42 U.S.C. § 7554 (a) (2012).
[5] 42 U.S.C. § 7412 (b)(2)(B) (2012).
[6] 33 U.S.C. § 1311 (b)(1)(A) (2010).
[7] 33 U.S.C. § 1314 (b)(1)(B) (2010).
[8] CWA, Pub. L. No. 107-377, Section 304(b)(2) (1972).

BOX 3-1
EPA Control Technology Categories

Best Practicable Control Technology Currently Available
"BPT is defined at Section 304(b)(1) of the [Clean Water Act (CWA)]. EPA sets Best Practicable Control Technology Currently Available (BPT) effluent limitations for conventional, toxic, and non-conventional pollutants. Section 304(a)(4) designates the following as conventional pollutants: biochemical oxygen demand (BOD5), total suspended solids, fecal coliform, pH, and any additional pollutants defined by the Administrator as conventional. The Administrator designated oil and grease as an additional conventional pollutant on July 30, 1979 (see 44 FR 44501)."

Best Conventional Pollutant Control Technology
"Best Conventional Pollutant Control Technology (BCT) is defined at Section 304(b)(4) of the CWA. The 1977 amendments to the CWA required EPA to identify effluent reduction levels for conventional pollutants associated with BCT for discharges from existing industrial point sources. In addition to the other factors specified in section 304(b)(4)(B), the CWA requires that EPA establish BCT limitations after consideration of a two part "cost-reasonableness" test. EPA explained its methodology for the development of BCT limitations in a *Federal Register* notice."

Best Available Technology Economically Achievable
"Best Available Technology Economically Achievable (BAT) is defined at Section 304(b)(2) of the CWA. In general, Best Available Technology Economically Achievable (BAT) represents the best available economically achievable performance of plants in the industrial subcategory or category. The factors considered in assessing BAT include the cost of achieving BAT effluent reductions, the age of equipment and facilities involved, the process employed, potential process changes, non-water quality environmental impacts, including energy requirements and other such factors as the EPA Administrator deems appropriate. EPA retains considerable discretion in assigning the weight according to these factors. BAT limitations may be based on effluent reductions attainable through changes in a facility's processes and operations. Where existing performance is uniformly inadequate, BAT may reflect a higher level of performance than is currently being achieved within a particular subcategory based on technology transferred from a different subcategory or category. BAT may be based upon process changes or internal controls, even when these technologies are not common industry practice."

SOURCE: EPA, 2012.

performance of plants in the industrial subcategory or category. If the variability of technologies within a category is too large, EPA can subdivide an industrial sector into more narrowly defined categories in order to examine available technologies in a more granular way.

Under the Safe Drinking Water Act (SDWA),[9] EPA must determine a nonenforceable maximum contaminant level goal (MCLG) for contaminants solely on the basis of human health risks.[10] In contrast, the enforceable drinking-water standard promulgated under the SDWA—termed the maximum contaminant level (MCL)—is the "level that may be achieved with the use of the best available technology, treatment techniques, and other means that EPA finds are available (after examination for efficiency under field conditions, not solely under laboratory conditions), taking cost into consideration."[11] Thus, EPA must consider technological and economic factors when setting the enforceable drinking water standards (the MCLs) but not the MCLGs.

Uncertainties in Technology Assessments

There is inherent uncertainty in the analyses of both current and future control technologies. When assessing current technologies to establish a BAT, for example, EPA must consider such parameters as the cost of achieving BAT effluent reductions, the age of the equipment and facilities involved, the process employed, potential process changes, and non-water-quality environmental impacts. There may be only limited data available about any or all of those parameters; for instance, EPA might have the facility age for only a subset of facilities within a sector or subcategory of a sector. There can also be variability in those parameters within a sector or subcategory of a sector, which contributes to the uncertainty in EPA's decisions. Using current technology for rulemaking, however, ignores that uncertainty, and it can lead to an underestimate of the level of control technologies that could be implemented. For example, mobile source control technology has steadily improved over the past 30 years. A 2005 review of technology innovation included information about how the technologies for catalyst and fuels preparation continued to improve over time with progressively more stringent tail pipe emission standards (ICF Consulting, 2005).

---

[9] SDWA, Pub. L. No. 93-523.

[10] When the statutory framework requires that decisions be made solely on the basis of health effects, the focus is likely to be on protecting maximally exposed or sensitive individuals. For such decisions there is no opportunity for other factors (for example, cost of mitigation, loss to property, loss of employment, and social factors) to influence the selection of management options.

[11] SDWA, Pub. L. No. 93-523.

The challenge is even greater when trying to predict what technologies might be available or in widespread use in the future and also to predict the effectiveness and costs of those technologies. For those predictions, analysts not only must estimate what technologies are currently available and the costs and amounts of emission reductions associated with those technologies, but also must model or somehow predict future developments in control technologies. The rate of innovation varies among sectors (Pavitt, 1984). An additional challenge is that past innovation curves might not reflect future rates of innovation. As businesses begin to implement technologies, there is a learning force at play that changes the rate of innovation. A regulation itself can also lead to an increase in the market size anticipated for a given technology, which provides a stimulus for investment in research and development for the technology. In other words, having regulations in place or on the horizon can lead to innovation in control technologies. In some sectors different businesses or entities will compete based on how efficiently they achieve a performance standard. This stimulates innovation and, as discussed later in this chapter, some analyses suggest that the innovation leads to costs that turn out to be less than estimated at the time of rulemaking (Morgenstern, 1997, 2011). Because there has been little study of innovation rates and the effects that EPA's regulations have on those innovation rates, there are insufficient data with which to develop models that account for those effects. Learning how technologies develop over time and how EPA's regulatory decisions affect that development could help improve the agency's models of technological factors and decrease the uncertainty in agency decisions.

EPA sometimes has the statutory authority, or the precedent, to consider both available technologies and technologies that are anticipated to be available. There are greater inherent unknowns associated with anticipating technologies—which in some cases could be considered deep uncertainty. In some cases, however, the agency will select a current technology but build into the rule a review or evaluation at a future date to update the rule as technology advances (see Chapter 2 for more discussion). An alternative to this approach is the development of performance standards rather than specific technologies.

Decision in the Face of Technology Uncertainties: EPA's Highway Heavy-Duty Engine Rule

In 1996 EPA was faced with uncertainties surrounding the technologies available to control emissions from highway heavy-duty engines. When considering regulations to control emissions of air pollution from highway heavy-duty engines (Control of emissions of air pollution from highway heavy-duty engines, 1996), EPA recognized that high concentrations of sulfur in diesel fuel led to corrosion caused by emission-control technologies that employ exhaust gas recirculation and that these sulfur concentrations were a major limiting factor to implementing emission-control technology. In its rulemaking EPA was faced with uncertainty about when and by what extent the concentrations of sulfur in fuel would decrease and about how low the concentration of sulfur would have to be in order not to affect emission-control technologies adversely.

Given those uncertainties, EPA took a more adaptive strategy to rulemaking. Rather than establishing a standard for emissions immediately, it established a timeline for future implementation of a standard and for evaluation of the effects of sulfur on control technologies in the interim. Specifically, in its rule EPA stated that "fuel changes could reduce the amount of emission control necessary for the engine, but . . . are probably not necessary to meet the proposed standards. However, this remains an area of uncertainty and is one of the issues which would be addressed further in the proposed 1999 review of the feasibility of the standard" (Control of emissions of air pollution from highway heavy-duty engines, 1996, p. 33455). In other words, EPA identified the potential effects of changing fuel composition as an uncertainty in the sense that it was not known to what extent decreasing the sulfur content and other changes in diesel fuel would affect the ability meet proposed emission standards, and the agency indicated that it would further evaluate those effects in a 1999 review of the standard. EPA's subsequent July 2000 document, *Regulatory Impact Analysis: Control of Emissions of Air Pollution from Highway Heavy-Duty Engines* (EPA, 2000c), reported that much lower concentrations of sulfur dioxide (a maximum of 30 ppm compared to then concentrations up to 500 ppm) would limit the corrosion caused by exhaust gas recirculation emission-control technologies.

In a separate rule EPA later required a significant reduction in the amount of sulfur in diesel fuels beginning with the 2007 model year. During the decade between when the diesel engine standards were set in the 1990s and when the standards took effect in 2007, EPA developed companion regulations to change fuel composition and implemented demonstration programs, while engine manufacturers developed innovative emission controls as they were preparing for the regulatory change. EPA was faced with an uncertainty about how sulfur content affects emission controls and, therefore, allowed the industry time for additional research, development, and commercialization of control technologies before having a regulation become effective. Such an approach allowed for the evaluation of recent technological innovations and updated standards that reflected those innovations. Scheduling such reviews of standards several years in the future can also motivate research and development.

ECONOMICS

As discussed above, a number of the statutes and executive orders[12] under which EPA operates require it to consider economics and economic analyses in its regulatory decisions. As described in EPA's *Guidelines for Preparing Economic Analysis* (EPA, 2010), economic analyses combine various types of information, including information from the assessments of other factors discussed in this and the previous chapter, to provide "a means to organize information and to comprehensively assess alternative actions and their consequences" (pp. 1–2). The analyses inform decision makers of the costs associated with the various risks, the benefits of reducing those risks, the costs associated with risk mitigation or remediation options, and the distributional effects (see Box 3-2 for EPA's definitions related to economic analysis). As with the other factors that affect EPA's decisions, economic analyses have uncertainties. Those uncertainties contribute to the overall uncertainty in a decision, and EPA should consider them its decision-making process.

This section discusses economic analysis and its uncertainty. The section begins with a brief overview of economic analysis in the regulatory setting. A number of texts and reports describe the use of economics and economic analyses in decision making in general (Gold et al., 1996; IOM, 2006; Sloan and Hsieh, 2012) and in environmental decision making in particular (Atkinson and Mourato, 2008; EPA, 2010; Pearce et al., 2006); the reader is referred to those sources for more detailed discussions. The committee then describes the uncertainties associated with those analyses, using examples to illustrate how EPA has evaluated and characterized uncertainty in economic analyses, followed by a discussion of the assessment of those uncertainties and reporting of those uncertainties.

Economic Analysis Approaches

Two of the main types of economic analysis are benefit–cost analysis (BCA; also called cost–benefit analysis) and cost-effectiveness analyses (CEA) (see Box 3-2 for definitions). In the context of environmental

---

[12] For example, Executive Order 12866 requires analyses of the costs and benefits of "significant regulatory actions." A "significant regulatory action" is defined as "any regulatory action that is likely to result in a rule that may: (1) Have an annual effect on the economy of $100 million or more or adversely affect in a material way the economy, a sector of the economy, productivity, competition, jobs, the environment, public health or safety, or State, local, or tribal governments or communities; (2) Create a serious inconsistency or otherwise interfere with an action taken or planned by another agency; (3) Materially alter the budgetary impact of entitlements, grants, user fees, or loan programs or the rights and obligations of recipients thereof; or (4) Raise novel legal or policy issues arising out of legal mandates, the President's priorities, or the principles set forth in this Executive order."

> **BOX 3-2**
> **Definitions of Select Terms Used in Economic Analyses as Defined in *Guidelines for Preparing Economic Analysis***
>
> **Benefit–cost analysis (BCA)**
> A BCA evaluates the favorable effects of policy actions and the associated opportunity costs of those actions. It answers the question of whether the benefits are sufficient for the gainers to potentially compensate the losers, leaving everyone at least as well off as before the policy. The calculation of net benefits helps ascertain the economic efficiency of a regulation.
>
> **Benefit–cost ratio**
> A benefit–cost ratio is the ratio of the net present value (NPV) of benefits associated with a project or proposal relative to the NPV of the costs of the project or proposal. The ratio indicates the benefits expected for each dollar of costs. Note that this ratio is not an indicator of the magnitude of net benefits. Two projects with the same benefit–cost ratio can have vastly different estimates of benefits and costs.
>
> **Benefits**
> Benefits are the favorable effects society gains due to a policy or action. Economists define benefits by focusing on changes in individual well-being, referred to as welfare or utility. Willingness to pay is the preferred measure of these changes, as it theoretically provides a full accounting of individual preferences across tradeoffs between income and the favorable effects.
>
> **Cost-effectiveness analysis (CEA)**
> CEA examines the costs associated with obtaining an additional unit of an environmental outcome. It is designed to identify the least expensive way of achieving a given environmental quality target or the way of achieving the greatest improvement in some environmental target for a given expenditure of resources.

regulations, the overall objective of both BCA and CEA is to compare different regulatory options,[13] different combinations of regulatory options, or the value of different regulatory options.[14] In BCA both benefits and costs are expressed in monetary units, whereas CEA is intended to identify

---

[13] Potential regulatory options include the option of taking no regulatory action.

[14] That comparison, even in cost-effectiveness analysis, requires characterizing value in a common unit; one convenient measure of common value is money (Gold et al., 1996; Sloan and Hsieh, 2012).

> **Costs**
> Costs are the dollar values of resources needed to produce a good or service; once allocated, these resources are not available for use elsewhere. *Private costs* are the costs that the buyer of a good or service pays the seller. *Social costs*, also called *externalities*, are the costs that people other than the buyers are forced to pay, often through nonpecuniary means, as a result of a transaction. The bearers of social costs can be either particular individuals or society at large.
>
> **Distributional analysis**
> Distributional analysis assesses changes in social welfare by examining the effects of a regulation across different subpopulations and entities. Two types of distributional analyses are the economic impact analysis (EIA) and the equity assessment.
>
> **Economic impact analysis (EIA)**
> An EIA examines the distribution of monetized effects of a policy, such as changes in industry profitability or in government revenues, as well as nonmonetized effects, such as increases in unemployment rates or numbers of plant closures.
>
> **Social cost**
> From a regulatory standpoint, social cost represents the total burden a regulation will impose on the economy. It can be defined as the sum of all opportunity costs incurred as a result of the regulation. These opportunity costs consist of the value lost to society of all the goods and services that will not be produced and consumed if firms comply with the regulation and reallocate resources away from production activities and towards pollution abatement. To be complete, an estimate of social cost should include both the opportunity costs of current consumption that will be foregone as a result of the regulation and also the losses that may result if the regulation reduces capital investment and thus future consumption.
>
> **Total cost**
> Total cost is defined as the sum of all costs associated with a given activity.
>
> SOURCE: EPA, 2010.

the most effective use of resources without requiring the monetization of all benefits or costs.

In 2003 the Office of Management and Budget (OMB) issued guidance on the development of regulatory analyses and the use of BCA and CEA. In that guidance OMB states that "major rulemaking should be supported by both types of analysis wherever possible" (OMB, 2003, p. 9). In 2011 EPA issued guidelines for economic analyses (EPA, 2010). In contrast to the OMB guidance, EPA's guidelines focus on conducting BCAs for economic analyses, although they mention that CEA can be used to help identify the least costly approach to achieving a specific goal.

BCAs are used to determine whether the benefits of a particular regulatory option justify its costs. When making decisions using a strict benefit–cost rule, a regulatory agency will adopt only those projects or implement only those regulations for which the present value of the net benefit (benefit minus cost) is non-negative or the benefit–cost ratio (present value of benefits divided by present value of cost) is one or greater. Under that strict rule, all projects for which the net benefit is negative or the benefit-to-cost ratio is less than one would be rejected. Alternatively, BCA can be used to rank regulatory options by the size of their net benefits; regulators can then choose the option or options with the largest net benefit. That approach, however, places small-scale projects at a disadvantage relative to larger ones. To account for project size, some economic analysts prefer to use a benefit-to-cost (B/C) ratio; if the B/C ratio is greater than one, the project is accepted, and otherwise it is rejected.

Description of Uncertainties in Economic Analyses

Uncertainty in economic analyses can stem from determining what costs and benefits should be including in the analyses, in the estimates of the costs and benefits themselves, and in adjusting the costs and benefits to reflect that they will occur in the future (that is, discounting).

The outcome of an economic analysis can vary greatly depending on the boundaries of that analysis, that is, on what is included in the analysis (Meltzer, 1997). Analyses can include mental health care costs and other health care costs, costs from lost employment, and a variety of other costs, such as the costs from the increased domestic violence that results from lost employment. In its guidelines, EPA (2010) considers *social cost*—which is described as the "total burden a regulation will impose on the economy" and "the sum of all opportunity costs incurred as a result of the regulation" (pp. xiv–xv)—to be the most comprehensive and appropriate measure of cost for a BCA. "Opportunity costs consist of the value lost to society of all the goods and services that will not be produced and consumed if firms comply with the regulation and reallocate resources away from production activities and toward pollution abatement" (p. xv). Social cost is narrower than the "total cost," which is considered in a number of regulatory impact analyses and includes costs beyond the social costs.

At the time of an economic analysis, the mean estimated values of anticipated benefits and costs and the corresponding net benefit or the benefit–cost ratio might indicate that a project is worth undertaking or that a rule is worth adopting, but after implementation the actual benefits and costs seen in retrospect can differ considerably from the estimated mean values because of uncertainties at the time of the analysis. For example, if there is an unanticipated increase in energy prices after a decision, the actual cost of a decision could exceed the anticipated cost, and what was thought to be a project of positive net benefit may turn out to have a substantial negative net benefit. That uncertainty, however, is rarely detailed in a BCA or CEA, let alone in the rationale for the decisions that use the BCA or CEA (NRC, 2002a).

We discuss below the uncertainties associated with the estimates of costs—categorized broadly as compliance costs and the costs imposed across the entire economy—and benefits as well as the uncertainties related to discounting those costs and benefits.

Cost Analysis

Compliance Costs

Compliance costs are the costs incurred in complying with a proposed regulatory rule, and they include those costs incurred by parties complying with the regulations (for example, the costs to install emission-control technologies in an industrial facility). These compliance costs are borne by regulated entities, of which many, if not most, are in the private sector (EPA, 2010). Additional compliance costs include the government's costs to monitor and enforce the rule and, more generally, activities to ensure compliance (Harrington et al., 1999). The increased costs borne by regulated entities and public agencies charged with enforcement may also be associated with increased costs borne by other private-sector entities and public agencies at different levels of government. An increase in enforcement effort will generally raise compliance rates and, subsequently, the costs borne by the regulated parties (Shimshack and Ward, 2005, 2008). Some research indicates, however, that the costs of "government administration of environmental statues and regulations" are "rarely considered in regulatory cost estimates" (Harrington et al., 1999, p. 9).

Measuring the capital costs and operating costs incurred by private parties in order to estimate compliance costs can be a difficult task for agencies with the responsibility of implementing a regulation. For example, under a cost-of-service regulation, an agency sets a price per unit of output and must gauge whether the regulated price it sets is sufficient to yield an adequate return to the regulated entity. At the same time, the agency must ensure that the price is not set so high that returns are excessive (Breyer, 1995).

Compliance costs are often estimated using engineering models, with the models often being based on expert opinions of the relationships between input use and outputs for a particular industry or application within an industry. It is not always made clear, however, from what settings the estimates were drawn. There may often be substantial variability in compliance costs depending on the characteristics of the regulated entity, including its scale and the age of its plants and equipment. For example, a plant may be so old that it would be replaced anyway, and the newer plant's design might already incorporate a technology that is consistent with the proposed rule. In other cases, the plant might be relatively new and at a much earlier point in the company's capital replacement cycle. In still other cases, if the cost of compliance is too great, facilities may be closed. It is also possible that, unknown to the EPA, particular facilities may have already been targeted for closure by company management even without the arrival of the new environmental requirements. A lack of such information is a common problem in the area of regulation and adds to the total uncertainty. EPA typically conducts surveys of facilities to determine the different types of technologies that are in place (EPA, 1995, 2004b). Cost estimates are usually based on estimates of the changes that would be required in the different types of facilities to comply with the standards.

EPA's decision documents rarely present a range of costs that represents the uncertainty in estimates of engineering costs,[15] and often it is not clear what assumptions underlay the computation of those cost estimates. For example, the summary of a 2000 regulatory impact analysis for arsenic in drinking water included tables listing the monetized health benefits from avoided cases of bladder and lung cancers and containing estimated compliance costs (EPA, 2000b). The table with the monetized benefits contains lower and upper estimates of benefits, which were based on the lower and the upper estimates of bladder cases avoided. No estimates that took sources of uncertainty other than human health risks into account are displayed. Estimates of the costs were provided in the summary table for two discount rates (3 percent and 7 percent), for two different plant categories, and for four different maximum contaminant levels. No analysis of other factors was displayed. As a result, the estimates did not reflect the overall variability in the cost of complying with the rule. A first step in dealing with this source of uncertainty would have been greater transparency in how the estimates were derived.

In a report that includes separate evaluations by different authors of three of EPA's regulatory impact analyses, Harrington et al. (2009) also cited a lack of consideration of the uncertainty in many parameters that affected the regulatory impact analysis.

The use of engineering models for estimating costs raises a number of other technical issues as well. Conceptually, the relevant compliance

---

[15] Engineering costs include the costs of purchasing, installing, and operating the technologies to meet a standard.

costs are marginal costs—that is, the incremental costs of complying with the rule—rather than average costs. There are issues of how joint costs or products are treated in the determination of engineering cost estimates. For example, the removal of one type of contaminant may be much less costly if industries have already installed treatment processes for other contaminants; similarly, if a control technology will decrease the emissions of a number of pollutants, it is difficult to know what portion of the costs of installing and maintaining that control technology should be attributed to regulating just one of those pollutants. In such cases the marginal cost of removing the contaminant can be substantially overstated if the other pollution control activities are not considered. EPA often estimates marginal costs and accounts for spillover effects, such as joint costs, in its analyses (EPA, 2011b; NAPEE, 2008). For example, EPA uses a model that accounts for the control of multiple pollutants (sulfur oxides, nitrogen oxides, directly emitted particulate matter, and carbon dioxide) in its regulatory impact analysis for mercury (EPA, 2011b).

There is also likely to be uncertainty concerning the number of households, firms, or systems (for example, water systems)[16] that may be affected by a rule and also concerning the methods that the regulated entities will use to comply with the rule. Uncertainty is even greater when EPA sets a national standard and agencies at a lower level of government, such as state agencies, implement the rule. In such instances, in addition to the issue of how firms will actually change to meet the new standards, there is additional uncertainty concerning how other units of government will implement the new standard. Once again, however, systematic inaccuracy is unlikely to occur, except in those cases in which a problem with compliance is known or anticipated.

Other sources of uncertainty that are sometimes relevant are the level of enforcement, the productivity of such enforcement efforts, and, subsequently, the compliance with the rule. Further increasing the uncertainty associated with compliance is the fact that in some cases lower levels of government enforce EPA's regulations. The simplest approach for dealing with such uncertainty is to assume complete compliance—in other words, 100 percent enforcement. EPA's guidelines recommend that when conducting regulatory impact analyses, analysts should, as a general rule, assume full compliance (100 percent) with EPA regulations (EPA, 2010). The guidelines recommend departure from using the "default" of full compliance only when there is sufficient data to calculate the true compliance rate (EPA, 2010). This level of enforcement may be higher than either the level that is socially optimal (that is, the one at which the marginal cost of

---

[16] See, e.g., *Federal Register*, November 22, 2001, p. 47, regarding an estimate of water systems affected by a proposal rule.

enforcement equals the marginal benefit of enforcement) or the level that occurs in practice. Estimates made with this assumption, therefore, should be considered to be high estimates of enforcement cost. Alternatively, EPA could estimate a range of costs using different percentages of firms complying with the laws. Whatever level of enforcement is assumed should be assumed throughout the analysis, including in the computation of benefits.

A few studies have compared the compliance costs estimated in regulatory impact analyses to estimates of actual compliance costs incurred after a regulation has been put into effect (Harrington, 2006; Harrington et al., 1999; OMB, 2005). Those comparisons indicate that compliance costs are often overestimated. OMB (2005) also concluded, however, that benefits are overestimated to a greater extent than costs, so that economic analyses typically predict that the performance of a regulation will be better than actually occurs. Harrington (2006) also found that although total costs were overestimated, unit costs were not, and he found "no bias in estimates of benefit–cost ratios." Regardless of which analysis is more accurate, all of those analyses demonstrate the uncertainty that is inherent in predictions of compliance costs (and benefits). And as discussed by Harrington (2006), these experiences also demonstrate the importance of conducting analyses after the implementation of a regulation (so-called ex post analyses) to evaluate and improve the methods used for predicting costs and benefits.

When a rule is to be implemented over a number of years, additional uncertainties arise. For example, input prices (that is, the price of inputs to a process, such as the price of low-sulfur coal) might vary with time. Moreover, as discussed earlier, the costs of the technological changes and equipment necessary for a plant to comply with the rule might change. For example, the promulgation of a rule on a national basis might increase the market size for an innovation that improves environmental quality. Such new technologies may be more productive in achieving a particular environmental goal, and in some cases the purchase prices of equipment incorporating the new technologies might be lower. At the time the rule is being considered, however, there is considerable uncertainty about how innovators will respond to the rule (that is, the amount of investment in research and development that will be forthcoming in response to promulgation of the rule), the yield from such investment, the time frame within which any yield will occur (that is, when the innovation will occur), and at which price the new technology will be marketed. A practical solution is to assume a worst case in which no innovation takes place, but such an assumption might underestimate the net benefit of innovation to the extent that there is an overestimation of the costs.

Estimates of compliance costs often must be made by estimating the number of facilities currently not in compliance with a proposed standard or rule, the magnitude by which those facilities would be out of compliance, and the current and future costs to bring those facilities into compliance under the new rule. All of those estimates are associated with an uncertainty that is difficult to accurately quantify. That uncertainty can, however, be qualitatively described, and potential ranges of costs can be used to provide decision makers with information on the effects of potential uncertainty on the estimates of the cost of different regulatory options.

Costs Imposed Economy Wide

The second broad category of costs consists of those costs imposed on other parties by increased prices (EPA, 2010). To the extent that prices increase, the quantities of output in other sectors are affected, which in turn affects the output for the economy as a whole (i.e., the gross domestic product). For example, if an environmental regulation increases the cost of coal mining, the price of coal is likely to increase, which in turn could lead to a decrease in coal consumption and an increase in the use of other energy sources. The increased price of coal would likely lead to an overall increase in energy costs, adding to the cost of manufacturing various products, which could in turn lower national production and employment.

When evaluating the broader costs of regulations, a distinction is often made between partial and general equilibrium analysis. Partial equilibrium analysis examines the effects of a regulatory change on a single firm. For example, a partial equilibrium analysis might assess the effect of a particular environmental regulation on the capital spending decisions of an individual firm. By contrast, general equilibrium analysis considers the effects of a regulatory change on all participants in a market or even in the economy as a whole. Individual sectors do not operate in a vacuum; if regulated firms increase the prices of their products, it may affect outputs (and prices) in other sectors. An analysis of those economy-wide effects, therefore, is often appropriate. Increased prices and reductions in output impose costs on society at large. However, if the potential effects of the regulatory rule are small or localized, there is little reason to assess its impacts on the economy as a whole.

It would be impractical to attempt to assess the economy-wide impacts of individual regulatory rules de novo. Instead it is necessary to employ models that have been developed for more general purposes. One useful tool that EPA has used for these purposes is a computable general equilibrium (CGE) model (EPA/RTI International, 2008). CGE models are a class of economic models that use actual economic data to estimate how an economy might react to changes in policy, technology, or other external factors. CGE models can also be used to compute the distributional impacts of regulatory changes. EPA has used a CGE model for a recent retrospective analysis of the benefits of the CAA (EPA, 2011a).

The starting point for an analysis with a CGE model is a set of assumptions about the impacts of the proposed rule on the output prices of the firms directly affected by the rule (RTI International, 2008). With the CGE model, the analyst computes prices and outputs of goods and services in various sectors and calculates the gross domestic product once the simulated economy has returned to a new equilibrium following implementation of the rule. CGE models are based on myriad assumptions about the underlying relationships among economic sectors—that is, about the substitutability of various goods and services in the economy. Those assumptions about these interrelationships, as well as the assumptions about price changes that are the essential inputs in the calculations, are sources of uncertainty in CGE models (RTI International, 2008).

There are also dynamic versions of CGE models that consider a broader range of longer-term effects, including technological changes, which have the potential to capture the long-term effects of regulatory rules on labor supply, savings, the growth of classes of inputs, and input productivity (RTI International, 2008). Structural changes in the economy occur over time as a result of a policy change. For example, a policy offering financial incentives to purchase energy-efficient appliances may lead to more demand for such appliances in the short run. In the long run, new appliances are developed because there is a greater financial incentive for firms to engage in research and development to develop new even more efficient projects. Because such long-term analyses rely on future projections, however, outcomes are far more uncertain than those obtained from use of static models.

One problem with CGE models is a lack of transparency. As discussed in the examples in Chapter 2, it is generally the case that few details are provided about either the baseline assumptions that were the starting points of the calculation or the key assumptions that went into development of the CGE model used for the economy-wide calculations. Thus it seems necessary for the decision maker to either accept the results at face value or to reject the exercise in its entirety.

Benefit Analysis

To project the increase in benefits attributable to an intervention, one must have a measure relating inputs to endpoints. In economics, that measure is termed a production function. The production function expresses a technical or scientific relationship, and the analyst typically obtains the production function from the scientific literature. For example, the production function could describe the effect that removing a carcinogen from the water supply has on the rates of particular forms of cancer. Given estimates of the production function parameters, it is possible to calculate the changes in health and other endpoints that are attributable to the intervention. Again, in a BCA the analyst must attach a pecuniary value to each endpoint.

Once the parameters of an assessment are established, estimating benefits requires establishing baseline values for the endpoints of interest, estimating the changes that would occur in those endpoints with different regulatory options (that is, the marginal effects of the policy or rule), and attaching a monetary value to a given endpoint (that is, valuing those endpoints). There are uncertainties inherent in each of these steps in a benefits assessment. An important source of uncertainty is the decision about which endpoints to include in the assessment, that is, determining the parameters the assessment will look at. Those three steps and the uncertainty associated with them are discussed below.

Establishing Baseline Values

The effect of a particular policy or rule on outcomes or endpoints of interest will depend, in part, on the original levels of what is being monitored. One exception is when the relationship between relevant effects of particular pollution levels is linear (EPA, 2010). Thus, the first step in calculating the benefit of a proposed rule is to calculate the values of a particular endpoint at a particular point in time when the change originates—that is, calculate the baseline benefits. For example, when estimating the benefits from a potential air pollution regulation, the rate of respiratory problems at a baseline point in time would be estimated. That point might be the time when the regulation is announced, if the industry is expected to reduce emissions in advance of the implementation of the law in order to prepare for the implementation, or at the time when the regulation is implemented if no changes in air pollution are anticipated in advance of implementation (NRC, 2002a).

Estimating the baseline involves a series of calculations, each with inherent uncertainties. The factors being measured could include human health as well as other factors with value to society, such as the preservation of specific species or habitats, atmospheric visibility, and pollution's effects on recreational use of various resources (EPA, 2010). The effects on those factors depend, in part, on the magnitude of pollutants which, in turn, depend on various output levels (EPA, 2010). For example, the effect of car emission regulations depends in part on the magnitude of the emissions from automobiles, which would depend in part on the number of cars on the road and the number of miles driven, and both of those outputs would vary by geographic location. The baseline variables are subject to scientific uncertainty as well as to uncertainties in activity levels in various sectors, which in turn can depend on exogenous factors (EPA, 2010). The number of miles driven, for instance, can depend on the price of gasoline, demographic changes, behavioral changes, and the existence of other regulatory rules and their levels of enforcement (EPA, 2010). In general, longer-range projections are subject to more uncertainty than shorter-range projections. Technological change can be a source of uncertainty in long-range projections (Moss and Schneider, 2000). For example, baseline air pollution is highly dependent on innovations in motor vehicle technology, such as the development of electric or hydrogen-powered automobiles. If baseline benefits are being estimated for a rule under consideration for automobile emissions and there are many years between when the proposed rule is developed and when it is implemented, uncertainty can come from trying to estimate improvements in the technology for electric cars and changes in consumer adoption of electric cars over time. Further complicating those estimates and creating more uncertainty is the fact that the proposed rule itself can encourage such innovation (EPA, 2010).

Multiple baselines can be used to indicate the range of potential baseline estimates, but they can make calculations very complex if all subsequent calculations need to account for multiple estimates of baseline benefits (EPA, 2010). Discussing baseline benefits with policy makers prior to estimating the baseline benefits can narrow the possible scenarios and decrease the number of baseline calculations that need to be performed. Many of the uncertainties associated with estimates of baseline benefits are deep uncertainties which will not be able to be resolved within the timeframe needed. In the face of such a high degree of uncertainty, transparency in the assumptions and analytic methods used to estimate baseline benefits is important in order to allow decision makers and stakeholders to understand how those benefits are estimated. If multiple baselines are not used, that fact should be clearly stated to make it clear that there is a source of uncertainty that is not being represented. Baseline estimates are the source of many errors in cost and benefit estimates (Harrington et al., 1999).

Marginal Effects of Policies and Rules

Once baseline benefits are estimated, the next step in estimating the overall benefits from a policy or rule is estimating the marginal effects of the policy or rule. In other words, one computes the anticipated changes from the baseline benefits that are attributable to the policy or rule in question and compares those estimates to estimates of the costs and benefits that would have occurred in the absence of the policy or rule. The same endpoints used in estimating the baseline benefits should be used to estimate the marginal effects of the rule.

The effects of a policy or rule on benefits are, in part, a function of the effects of the policy or rule on the exposures of a population to the harmful substance or substances being controlled by the policy or rule. Uncertainties about the effects of the rule on the exposure of a population to a harmful substance or substances come from uncertainties about rule enforcement and compliance, about the dose–response relationship, and about the time path of the response.

Valuing Endpoints

The final step in evaluating benefits is to attach a monetary value to each endpoint. Those endpoints could be related to the effects that a regulation had on human health or on such things as access to a park or clean air or community health. The monetary value placed on these things reflects society's maximum willingness to pay and is generally expressed per unit change in the endpoint—for example, how much society is willing to pay for each life-year saved or each day of hospitalization averted (EPA SAB, 2009; NRC, 2004). There can be large variability in how different endpoints are valued, which adds to the uncertainty in the economic analysis.

There are essentially two broad approaches for valuing the effects of policies: the revealed-preference approach and the stated-preference approach (Adamowicz et al., 1997; EPA SAB, 2009; Sloan and Hsieh, 2012; Williams, 1994). The advantages and disadvantages of those two methods are reviewed in detail elsewhere (EPA SAB, 2009; Freeman, 2003). The uncertainties in the approaches are discussed below.

Revealed-Preference Approach The revealed-preference approach bases valuations on actual decisions that people make, for example, the additional wage rate that compensates workers for taking a job with a higher fatality risk (Viscusi and Aldy, 2003). Revealed-preference studies are available for some, but by no means all, endpoints that need to be valued for environmental policy decisions (Boyd and Krupnick, 2009). Although there are many revealed preference–based studies of the value of life and the value of life-years, few analyses of morbidity and disability exist. Furthermore, many of the endpoints that are valued in benefit analyses for environmental policy decision making do not have inherent commercial or market value. For such endpoints, it is necessary to rely on valuations obtained from stated preference approaches (Yao and Kaval, 2011).

The major source of uncertainty is that there are few valid, reliable, and representative estimates of societal willingness to pay for reductions in morbidity and disability. Given the paucity of relevant studies, valuations tend to be based on a few studies of local populations that may not be nationally representative. It is not the existence of uncertainty in the valuation that is problematic, however, but rather it is when those uncertainties are ignored—that is, not quantified or somehow accounted for in decisions—that problems arise.

Furthermore, several potentially important dimensions of benefit are not valued at all. For example, in an economic analysis of a rule governing arsenic in drinking water, several factors that could influence the value of the benefits from removing arsenic from the drinking water—for example, not having to accept an involuntary risk of being exposed to arsenic in the water supply—are not currently given a value (EPA, 2000a). The Arsenic Benefits Review Panel of EPA's Science Advisory Board (EPA, 2001), in its review of EPA's economic analysis of the arsenic rule (EPA, 2000b), noted that "some people may value the existence of lower levels of arsenic in drinking water, possibly for psychological reasons (e.g., dread of being exposed)" (p. 3). That is, aside from the health risk of being exposed to arsenic, arsenic exposure has a variety of other costs. Having arsenic in the water supply can be a source of anxiety for those individuals directly exposed to it, for example, and the fact that some persons are exposed to arsenic in the water supply can be seen as unfair, a negative feature over and above any adverse health effects arsenic might have.

Stated-Preference Approach The stated-preference approach, also called contingent valuation, bases valuations on surveys designed to determine the willingness of a household to pay for a policy that will produce benefits for that household. Contingent valuation has been used to place value on nonmarket items, such as the worth of access to a park or of clean air or of community health. Many studies that use contingent valuation address ecological issues, such as the value of preserving bald eagles, wetlands, forests, or visibility at national parks (Breedlove, 1999), and they can identify priorities among various types of concerns, such as air quality; trash, illegal dumping, and abandoned housing; economic development; parks and surface water quality; and community health. The surveys are a mechanism for stakeholder input into the decision-making process and can be used when estimating the benefits of a regulatory decision related to human health. They are also a source of uncertainty in regulatory decisions, including from variability in the community's views and a lack of knowledge about that variability, and uncertainty in the techniques used to assess those views.

EPA's Science Advisory Board (SAB) has published a number of reports on ecological valuation methods, how EPA could apply those methods in its decision-making process, and the value that communities place on ecology. In 2006 the board held a workshop on ecological risk assessment and environmental decision making, and the following year it issued a report on that subject (EPA, 2007). The report demonstrates the recognition by the SAB that EPA needs a broader approach to ecological risk assessment and decision making. In particular, the SAB stated, "Local and regional regulatory processes are conditioned by community values and economic objectives as well as by ecological conditions. Therefore, aligning the decision and the supporting risk and economic analyses with 'what matters to people' is essential to achieve acceptable risk solutions that can be easily and effectively communicated to the public" (EPA SAB, 2007, p. ii). To achieve that alignment the SAB recommends that EPA "increase its understanding of and capacity to utilize ecosystem valuation methods in conjunction with such decisions" (EPA SAB, 2007, p. ii).

The SAB has extended this work in a report on valuing ecological systems and services which includes a summary of the contingent valuation literature (EPA, 2009). Some of the relevant recommendations on implementation involve seeking information about public concerns and needs using a variety of methods, including interactive processes to elicit public values; describing the valuation measures in terms that are meaningful and understandable to the public; and providing information to decision makers about the level of uncertainty in the valuation efforts.

The Superfund site cleanup program provides EPA with the opportunity and mandate to consider public values, and it is one for which EPA has evaluated community concerns. For example, the *Hudson River PCBs Superfund Site New York Record of Decision* (EPA, 2002a) details community concerns about proposed remediation activities. Community concerns included traffic, noise, construction lighting, air quality, odor, aesthetics, and a loss of recreational activities on the river. In 2004 EPA released a document on performance standards for quality-of-life concerns (EPA, 2004a) that addressed odor, noise, construction lighting, and navigation. Other quality-of-life considerations (aesthetics, road traffic, and recreational activities on the river) were reviewed as part of the development of performance standards, but it was determined that they did not need a performance standard. No quality-of-life standard for water quality was issued because other standards and regulatory requirements dealt with water quality.

While the quality-of-life issues were not part of the decision by EPA to remediate the Hudson River site per se, these issues were considered in decisions about the day-to-day process for remediation. For example, EPA assessed the levels of noise from remediation activities. The agency stated that it "has determined that the noise associated with construction and continuous operation of the sediment process/transfer facilities and hydraulic and mechanical dredging operations is not expected to be a significant concern" (EPA, 2004a, pp. 4-3, 4-4). The basis for that assessment, however, is not clear in the record of decision, the responsiveness summary, or the white papers that make up the lengthy documentation for the Hudson River cleanup (EPA, 2002b).

Under the Superfund Law, the Department of the Interior and the National Oceanic and Atmospheric Administration have used contingent valuation to calculate damages to natural resources (Lipton et al., 1995).

Contingent valuation has also been used, for example, to assess people's preferences among three endangered species (Wallmo and Lew, 2011).

In 2005 the Irish Environmental Protection Agency undertook a nationwide survey of citizen's views on litter, illegal dumping, and the remediation of illegal dumpsites (IEPA, 2006). The survey examined the level of concern, demand for greater enforcement, and willingness to pay among citizens. The survey results suggested that waste management was by far the most important environmental issue facing that country—56 percent of the 1,500 respondents chose that answer compared to, for example, the 2 percent who identified factory emissions or the 9 percent who identified planning and green spaces as the most important environmental issue. While the committee did not identify or explore how the Irish agency used this information, the survey results shed light on how the people of that nation value environmental issues and on how those values can be assessed.

Tools and surveys are available for use in contingent valuation. *Managing Risks to the Public: Appraisal Guidance* (HM Treasury, 2005) presents a concern assessment tool that provides "a framework for understanding people's concerns in order that they can be considered in policy development and in the development of related consultation agreements and communication strategies" (p. 33). The framework is based on findings related to risk perception (Fischhoff, 1995; Slovic, 1987) and includes questions about familiarity and experience with a hazard, understanding of cause and effect, the fairness of the distribution of risks and benefits, the fear or dread of a risk, and the trust that people have in the agency (HM Treasury, 2005).

Survey techniques and contingent valuation have both critics and supporters (Diamond and Hausman, 1994; Epstein, 2003; Hanemann, 1994; Portney, 1994). For example, Yeager and colleagues (2011) demonstrated the sensitivity of responses to survey question design and argued that caution should be exercised when interpreting surveys. There is also evidence that responses to such surveys reflect what people are willing to pay for something in general, but not necessarily what they are willing to pay for the specific item they are being questioned about—an observation called the list paradox. For example, once a person has indicated a willingness to pay $5 for clean air, if asked about valuing another item, such as Yosemite National Park or the other 57 national parks, he or she will also often say $5. The "willingness" to pay $5 is actually a signal that the person cares about the environment, and it does not indicate that the person, upon reflection, would actually be willing to pay $5. In addition, Kahneman (2011) pointed out that because people tend to think only about what is in front of them and neglect the opportunity costs, people make better decisions if they think more broadly. Showing people a full list of problems shows the opportunity costs clearly and can help balance the responses. Brookshire and Coursey (1987) demonstrated that an individual's valuation can change as he or she become more familiar with the survey methods and "the degree in which values are measured in a market or nonmarket environment."

An issue that is often not addressed directly is the fact that arguments about uncertainty are often a proxy for other public concerns. As discussed further in Chapters 5 and 6, when regulators, stakeholders, and the public work together in a decision-making process and disagreements occur, consideration should be given to what might be motivating the public's concerns, questions, and oppositions to decisions. Concerns about a decision and its uncertainties sometimes stem not from the specific uncertainties, but rather from dissatisfaction about the process that led to the decision (Covello et al., 2004; Hance et al., 1988). The scientific and technical disagreements can be more readily identified and resolved if people have an opportunity to air their concerns and if everyone has a common understanding of what the regulatory options and legal context are, of the science and the economics, and of what technologies are available (IOM, 2012). Mistrust of government is often at the root of what may appear at first to be scientific or technical disagreements.[17] For example, Santos and Edwards (1990) explored the underlying issues in disagreements about citing a nuclear power plant. They concluded, "The bottom line was that the public did not trust the government's and nuclear industry's ability to control human error. Unfortunately, there appears to have been no direct dialogue about this. Instead, the risk assessment itself [for the nuclear power plant siting] became the target for challenge" (p. 60).

Discounting

Current costs and benefits accrued in the present are typically considered to be worth more than costs and benefits accrued in the future, based on the fact that individuals generally prefer to enjoy benefits now rather than later (see, for example, Gold et al., 1996). So when assessing the costs and benefits associated with an environmental regulation, future costs and benefits are discounted compared to current costs and benefits. Because the discount rate accounts for the preference for current benefits, there is no "correct" discounting rate, which creates uncertainty about how to select the appropriate discount rate and which also adds to the variability in the cost and benefit assessments. Differences of opinions about the appropriate discounting rate are unlikely to be resolved, moving it toward being a deep uncertainty. Therefore, it is less important to choose a discount rate than it is to provide information about how different discount rates affect the analysis of a regulation.

---

[17] Social trust is discussed further in Chapter 6.

In practice, discount rates of 3 and 7 percent are in widespread use for projects of only a few decades in duration (EPA/RTI International, 2008). Some research indicates that long-term rates are less than short-term rates, and therefore, for projects affecting more than one generation, such as those involving climate change, a lower discount rate or varying discount rates that are given by a schedule may be appropriate (Newell and Pizer, 2003, 2004). These "default" assumptions indicate the uncertainty inherent in this component of economic analyses. The most direct way of accounting for such uncertainties is to evaluate the benefits versus the costs of a particular decision using alternative discount rates and to judge how estimates of net benefits or benefit-to-cost ratios are affected and if they are affected sufficiently to affect the decision about project or rule adoption. As the committee indicated in Chapter 1, a sole focus on uncertainties in the risk assessment ignores possibly even greater uncertainties in key factors in decision making.

Assessment of Uncertainties in Economic Analyses

There are several methods for accounting for the uncertainty in economic analyses in a decision, ranging from the very simple to the highly complex. Simple solutions include inflating costs, deflating benefits, or both, and establishing a threshold for a regulatory action, such as only adopting those projects for which the projected benefit is 1.25 times the cost. That approach is similar to the use of defaults in estimating health risks and, similarly, has inherent uncertainties that EPA should consider when weighing economic factors. An example of a more complex approach to dealing with uncertainty in economic analyses can be found in OMB's guidelines on the development of regulatory analysis (OMB, 2003). Those guidelines recommend specifying the entire probability distribution of benefits and costs by, for example, using Monte Carlo simulations to assess the degree of uncertainty in estimates of benefits, costs, and benefit–cost ratios (OMB, 2003). Monte Carlo simulations are widely used in business applications and also, as mentioned in Chapter 2, for estimating human health risks (Marshall et al., 2009; Thompson et al., 1992).

People's decisions about when they can afford to retire provide an example of how Monte Carlo analyses can aid decision making. By running multiple simulations using typical future consumption levels and the probability of different rates of return on investments and longevity, both of which are considered to be uncertain, Monte Carlo simulations can estimate the probability that an individual near retirement will outlive his or her assets, given the person's initial asset mix. For a Monte Carlo simulation of a BCA or CEA, the input would be the range of potential benefits and the likelihood that each benefit would turn out to be the "true benefit" that is actually seen. Similarly, the range of potential costs and the likelihood that each cost is the "true" or "actual" cost would be used to determine the probable cost. OMB's guidelines (OMB, 2003) and the NRC report *Estimating the Public Health Benefits of Proposed Air Pollution Regulations* (NRC, 2002b) both recommend probabilistic modeling of uncertainty for BCAs and CEAs. Although such models are not commonplace in EPA's estimates of benefits and costs (Jaffee and Stavins, 2007), the agency has used them at times. In a Monte Carlo analysis of benefits and costs, one calculates the probability that the actual net benefit will be negative or else that the net benefit will be less than a prespecified threshold amount.

Reporting of Uncertainty

A common problem when decision makers are faced with uncertainty related to a decision is how the uncertainty should be presented to the decision makers. Estimates of uncertainty in BCA and CEA are often presented as aggregate numbers. Aggregates, however, do not by themselves help a decision maker identify the individual sources of uncertainty, but rather they indicate the overall level of certainty in the analyses. Knowing the specific sources of uncertainty can be as important as—if not more important than—documenting uncertainties in the aggregate. If the sources of uncertainty are identified, it may be possible to decrease the uncertainty by conducting further research or to refine the policy option in order to reduce the impact of the uncertainty. For example, if it is known that the uncertainty in a benefit valuation is due to heterogeneity, then the identification of the groups or stakeholders who would bear the burden of a higher-than-anticipated cost or of a lower-than-anticipated benefit because of the uncertainty might allow decision makers to design the initial proposed regulations to address or cope with those potential outcomes in advance. Although such analyses are possible, they are rarely done for BCAs and CEAs (NRC, 2002a). If the individual sources of uncertainty that contribute to the overall uncertainty can be determined, uncertainty analyses in BCAs and CEAs can incorporate graphic representations—such as Tornado plots where the relative importance of the different sources of uncertainty are displayed sequentially—to provide an easily interpretable graphic display of the sources of uncertainty (Krupnick et al., 2006).

Arsenic in Drinking Water

As discussed in Chapter 2, in 2001 EPA evaluated a recently enacted standard that would decrease the allowable concentration of arsenic in drinking water. When estimating the cost associated with implementing the proposed arsenic rule, EPA used a Monte Carlo simulation to forecast "a distribution of costs around the mean compliance cost expected for each system size category" (EPA, 2000a).[18] The uncertainty analysis included only treatment costs, and it assumed a single commercial discount rate of 5 percent. In reviewing EPA's cost estimates for the arsenic rule, the Arsenic Cost Working Group of the National Drinking Water Advisory Council noted that the "value of existing national cost estimates is now limited by the large uncertainty associated with the estimated outcomes" and recommended that the EPA "clearly explain the limitations of each estimate and quantify the uncertainty associated with the Arsenic Rule estimates" (Arsenic Cost Working Group to the National Drinking Water Advisory Council, 2001, p. 2). The advisory council also noted the need for "a more representative methodology to assess compliance cost," noting the advantages of "an approach based on aggregated county, regional or state costs, coupled with extensive individual case analysis" (p. 2). Furthermore, the council recommended the use of a distribution of flows rather than the mean or median flow and noted the "significant uncertainty associated with EPDS [the number of entry points to the distribution system] determination" (p. 3), and it recommended how EPA should examine information given that uncertainty.

The relative lack or small amount of uncertainty analyses performed for the analysis of cost and benefits in the arsenic rule contrasts sharply with the extensive uncertainty analyses that were conducted for the human health risk estimates. EPA and the National Research Council conducted extensive analyses of the health data in order to estimate health risks, and that work included an extensive discussion of inherent uncertainties as well as quantitative assessments of how many of those uncertainties might affect the health risk estimates.

Some of those limitations were outlined by the Arsenic Rule Benefits Review Panel of EPA's SAB (EPA, 2000b). That panel noted that in the estimates for the number of cancer cases avoided there was nothing said about the uncertainty in the assumptions made about the lag time between the reduction of arsenic exposures and the reduction in risk. It also noted that the benefits and costs should be summarized in a manner that would indicate the variability in the benefits and costs associated with different sizes of water treatment facilities. It recommended that the age distribution of cancer cases avoided should be presented in order to allow readers to know the age distribution of those benefits, and it discussed the limitations of EPA's valuation of avoided cancer morbidity and mortality, recommending that the agency conduct more uncertainty analyses around those valuations, using sensitivity analyses or Monte Carlo analyses.

---

[18] In that report EPA noted that, "Historically, most drinking water regulatory impact analyses used point estimates to describe the average system-level costs" (EPA, 2000a, pp. 6–17).

OTHER FACTORS

A number of factors other than human health risks, technological availability, and economics affect EPA's decisions. Some of those factors and the uncertainty in estimates of them can be accounted for in estimates of human health risks (for example, the adverse health effects on sensitive populations such as infants and children) or in economic analyses (for example, the value that the public places on having access to recreational space is often accounted for in BCAs). Other factors—such as environmental justice, political climate, and public sentiment—and their uncertainty, however, are not taken into account in the analyses of human health risk, economics, and technological availability, despite their influence on decisions.

Executive Order 13563, issued in 2011, states that regulations are "to be based, to the extent feasible and consistent with law, on the open exchange of information and perspectives among State, local, and tribal officials, experts in relevant disciplines, affected stakeholders in the private sector, and the public as a whole."[19] As defined by EPA, public values "reflect broad attitudes of society about environmental risks and risk management" (EPA, 2000d, p. 52). Public values can be specific to a certain geographic area or can apply to the nation as a whole. As such, public values affect both local- and national-level decisions and can affect EPA's regulations. Several programs within EPA take community concerns into consideration, as required by either statute or executive order. The Resource Conservation and Recovery Act and the Comprehensive Environmental Response, Compensation, and Liability Act, commonly known as Superfund, address several issues related to public values.

A community, for example, has an interest when it is chosen as a location for a hazardous waste facility or when an existing facility is permitted. EPA provides guidance for industries and agencies working with communities likely to be affected by a hazardous waste site location; in that guidance the agency describes potential community concerns, such as concerns about the effects on quality of life, including concerns about preserving the community's use of space, enjoyment and value of property, and sense of belonging and security, as well as promoting the economically sound protection of resources (EPA, 2000e). The guidance does not provide advice on integrating those concerns into the decision-making process. Others define public participation as "the process by which public concerns, needs and values are incorporated into governmental and corporate decision making" (Creighton, 2005, p. 7). The broader aspect of public participation is recommended by others, including a National Research Council committee

---

[19] Exec. Order No. 13563. Improving regulation and regulatory review. 76 FR 3821 (January 21, 2011).

(NRC, 2008) and EPA's SAB Committee on Valuing the Protection of Ecological Systems and Services (EPA, 2009).[20]

The political climate can also affect EPA's decisions. Although EPA is a scientific regulatory agency, its decisions take place in a broader context that includes more than just the scientific issues. As described by EPA (2000d), political factors that can affect decisions are "based on the interactions among branches of the Federal government, with other Federal, state, and local government entities, and even with foreign governments; these may range from practices defined by Agency policy and political administrations through inquiries from members of Congress, special interest groups, or concerned citizens" (p. 52). Thus the agency is influenced by a complex set of forces, not the least of which are the values, priorities, and direction of the President. No decision is based absolutely or purely on scientific analyses. Regardless of statutory directives, agencies often have broad discretion about when to act in the face of a risk, and a decision of when to act is influenced by executive branch leadership, congressional intention and attention, and the possibility of judicial intervention.

Although social considerations such as environmental justice and the political climate affect EPA's decisions and there is uncertainty in those factors and how they influence decisions, there is seldom any discussion concerning just how those factors and their uncertainty affect a decision.

KEY FINDINGS

- EPA considers a number of factors in addition to human health risks when making regulatory decisions. EPA's legal authority and requirements predominately determine what factors it considers when making regulatory decisions.
- Uncertainty is present in the assessment of all of the factors EPA that considers when making regulatory decisions, including technology availability and economic factors. Those uncertainties, however, are rarely analyzed or explicitly accounted for in EPA's regulatory decisions. Similarly, factors such as public sentiment, environmental justice, and the political climate influence EPA's decisions, but the uncertainty in those factors is rarely accounted for in EPA's decisions. EPA also does not discuss the uncertainty in any of those factors in its decision documents as thoroughly as it does the uncertainty in human health risk estimates.
  o Although different estimates of technology availability are sometimes used, the uncertainties that stem from those estimates are

---

[20] Chapter 6 further discusses the importance of stakeholder engagement and communication approaches for that participation.

not often carried through to the final outputs in a regulatory impact assessment.
- Uncertainties in economic analyses are sometimes conducted, but they are not necessarily presented.
- Uncertainties in public sentiment and social factors, such as environmental justice, are rarely accounted for explicitly in decisions, and their effects are rarely discussed.
- The political climate can affect assessments of regulatory options. That effect of that climate on decisions is not always transparent, adding to the uncertainty in decisions.
• The methods and use of uncertainty analyses for technological, economic, and other factors are not as well established as for the uncertainties in human health risk estimates. The committee's review of the literature indicated that uncertainty analyses for economic analyses have been studied more than uncertainty analyses for the technological and social factors.

RECOMMENDATION 2
The U.S. Environmental Protection Agency (EPA) should develop methods to systematically describe and account for uncertainties in decision-relevant factors in addition to estimates of health risks—including technological and economic factors—in its decision-making process. When influential in a decision, those new methods should be subject to peer review.

RECOMMENDATION 3
Analysts and decision makers should describe in decision documents and other public communications uncertainties in cost–benefit analyses that are conducted, even if not required by statute for decision making, and the analyses should be described at levels that are appropriate for technical experts and non-experts.

RECOMMENDATION 4
The U.S. Environmental Protection Agency should fund research, conduct research, or both to evaluate the accuracy and predictive capabilities of past assessments of technologies and costs and benefits for rulemaking in order to improve future efforts. This research could be conducted by EPA staff or else by nongovernmental policy analysts, who might be less subject to biases. This research should be used as a learning tool for EPA to improve its analytic approaches to assessing technological feasibility.

RECOMMENDATION 5

The U.S. Environmental Protection Agency should continue to work with stakeholders, particularly the general public, in efforts to identify their values and concerns in order to determine which uncertainties in other factors, along with those in the health risk assessment, should be analyzed, factored into the decision-making process, and communicated.

RECOMMENDATION 6

The U.S. Environmental Protection Agency should fund or conduct methodological research on ways to measure public values. This could allow decision makers to systematically assess and better explain the role that public sentiment and other factors that are difficult to quantify play in the decision-making process.

REFERENCES

Adamowicz, W., J. Swait, P. Boxall, J. Louviere, and M. Williams. 1997. Perceptions versus objective measures of environmental quality in combined revealed and stated preference models of environmental valuation. *Journal of Environmental Economics and Management* 32(1):65–84.

Arsenic Cost Working Group to the National Drinking Water Advisory Council. 2001. *Report of the Arsenic Cost Working Group to the National Drinking Water Advisory Council.* Washington, DC: Environmental Protection Agency.

Atkinson, G., and S. Mourato. 2008. Environmental cost–benefit analysis. *Annual Review of Environment and Resources* 33:317–344.

Boyd, J., and A. Krupnick. 2009. *The definition and choice of environmental commodities for nonmarket valuation.* Washington, DC: Resources for the Future.

Breedlove, J. 1999. *Natural resources: Assessing non-market values through contingent valuation.* Washington, DC: Congressional Research Service.

Breyer, S. 1995. *Breaking the vicious circle: Toward effective risk regulation.* Cambridge, MA: Harvard University Press.

Brookshire, D. S., and D. L. Coursey. 1987. Measuring the value of a public good: An empirical comparison of elicitation procedures. *American Economic Review* 554–566.

Control of emissions of air pollution from highway heavy-duty engines. 1996. *Federal Register* 61(125):33421–33469.

Covello, V. T., D. B. McCallum, and M. T. Pavlova. 2004. *Effective risk communication: The role and responsibility of government and nongovernment organizations.* New York: Plenum Press.

Creighton, J. L. 2005. *The public participation handbook: Making better decisions through citizen involvement.* San Francisco, CA: Jossey-Bass.

Diamond, P. A., and J. A. Hausman. 1994. Contingent valuation: Is some number better than no number? *Journal of Economic Perspectives* 8(4):45–64.

EPA (U.S. Environmental Protection Agency). 1995. *Survey of control technologies for low concentration organic vapor gas streams.* Washington, DC: EPA.

———. 2000a. *Arsenic in Drinking Water Rule economic analysis.* Washington, DC: Office of Ground Water and Drinking Water, EPA.

———. 2000b. *Proposed Arsenic in Drinking Water Rule regulatory impact analysis.* Washington, DC: Office of Ground Water and Drinking Water, EPA.
———. 2000c. *Regulatory impact analysis: Control of emissions of air pollution from highway heavy-duty engines.* Washington, DC: PA.
———. 2000d. *Social aspects of siting RCRA hazardous waste facilities.* Washington, DC: EPA.
———. 2001. *Arsenic rule benefits analysis: An SAB review. A review by the Arsenic Rule Benefits Review Panel (ARBRP) of the U.S. EPA Science Advisory Board (SAB).* Washington, DC: EPA Science Advisory Board.
———. 2002a. Hudson River PCBs site New York record of decision. Washington, DC: Environmental Protection Agency. http://www.epa.gov/hudson/RecordofDecision-text.pdf (accessed November 21, 2012).
———. 2002b. Responsiveness summary: Hudson River PCBs site record of decision— White papers. Washington, DC: Environmental Protection Agency. http://www.epa.gov/hudson/d_rod.htm#response (accessed November 21, 2012).
———. 2004a. Hudson River PCBs Superfund site quality of life performance standards. Washington, DC: Environmental Protection Agency. http://www.epa.gov/hudson/quality_of_life_06_04/full_report.pdf (accessed November 21, 2012).
———. 2004b. *Survey of technologies for monitoring containment liners and covers.* Washington, DC: EPA.
———. 2007. *Advice to EPA on advancing the science and application of ecological risk assessment in environmental decision making: A report of the U.S. EPA Science Advisory Board.* Washington, DC: Environmental Protection Agency, Science Advisory Board.
———. 2009. *Valuing the protection of ecological systems and services: A report of the EPA Science Advisory Board.* Washington, DC: EPA Science Advisory Board.
———. 2010. Guidelines for preparing economic analyses. http://yosemite.epa.gov/ee/epa/eerm.nsf/vwAN/EE-0568-51.pdf/$file/EE-0568-51.pdf (accessed November 21, 2012).
———. 2011a. *The benefits and costs of the Clean Air Act from 1990 to 2020. Final report.* Washington, DC: Environmental Protection Agency, Office of Air and Radiation.
———. 2011b. *Regulatory impact analysis for the final mercury and air toxics standards.* Research Triangle, NC: Environmental Protection Agency, Office of Air Quality Planning and Standards, Health and Environmental Impacts Division.
———. 2012. *Water: Industry effluent guidelines.* Washington, DC: EPA. http://water.epa.gov/scitech/wastetech/guide/questions_index.cfm (accessed November 21, 2012).
EPA/RTI International. 2008. *EMPAX–CGE model documentation interim report.* Research Triangle Park, NC: EPA.
EPA SAB (EPA Science Advisory Board). 2007. Advice to the EPA. http://yosemite.epa.gov/sab/sabproduct.nsf/7140dc0e56eb148a8525737900043063/$file/sab-08-002.pdf (accessed November 21, 2012).
———. 2009. Valuing the protection of ecological systems and services. http://yosemite.epa.gov/sab/sabproduct.nsf/F3DB1F5C6EF90EE1852575C500589157/$File/EPA-SAB-09-012-unsigned.pdf (accessed November 21, 2012).
Epstein, R. A. 2003. The regrettable necessity of contingent valuation. *Journal of Cultural Economics* 27(3):259–274.
Fischhoff, B. 1995. Ranking risk. *Risk: Health, Safety, and Environment* 6:189–200.
Freeman, A. M. 2003. *The measurement of environmental and resource values: Theory and methods.* Washington, DC: Resources for the Future.
Gold, M. R., J. E. Siegel, L. B. Russell, and M. C. Weinstein. 1996. *Cost effectiveness in health and medicine.* New York: Oxford University Press.

Hance, B., C. Chess, and P. Sandman. 1988. *Improving dialogue with communities: A risk communication manual for government*. Trenton: New Jersey Department of Environmental Protection.
Hanemann, W. M. 1994. Valuing the environment through contingent valuation. *Journal of Economic Perspectives* 8(4):19–43.
Harrington, W. 2006. *Grading estimates of the benefits and costs of federal regulation*. Washington, DC: Resources for the Future.
Harrington, W., R. Morganstern, and P. Nelson. 1999. On the accuracy of regulatory cost estimates. http://www.rff.org/documents/RFF-DP-99-18.pdf (accessed November 21, 2012).
Harrington, W., L. Heinzerling, and R. D. Morgenstern. 2009. *Reforming regulatory impact analysis*. Washington, DC: Resources for the Future.
HM Treasury. 2005. *Managing risks to the public: Appraisal guidance*. London: Crown.
ICF Consulting. 2005. *The Clean Air Act amendments: Spurring innovation and growth while cleaning the air*. Fairfax, VA: ICF Consulting.
IEPA (Irish Environmental Protection Agency). 2006. *Public perceptions, attitudes and values on the environment—A national survey*. Wexford, Ireland: Irish Environmental Protection Agency.
IOM (Institute of Medicine). 2006. *Valuing health for regulatory cost-effectiveness analysis*. Washington, DC: The National Academies Press.
———. 2012. *Ethical and scientific issues in studying the safety of approved drugs*. Washington, DC: The National Academies Press.
Jaffee, J., and R. N. Stavins. 2007. On the value of formal assessment of uncertainty in regulatory analysis. *Regulation and Governance* 1:154–171.
Kahneman, D. 2011. *Thinking, fast and slow*. New York: Farrar, Straus, Giroux.
Krupnick, A., R. Morganstern, M. Batz, P. Nelson, D. Burtraw, J. Shih, and M. McWilliams. 2006. *Not a sure thing: Making regulatory choices under uncertainty*. Washington, DC: Resources for the Future.
Lipton, D. W., K. Wellman, I. C. Sheifer, and R. F. Weiher. 1995. *Economic valuation of natural resources: A handbook for coastal resource policymakers*. Silver Spring, MD: U.S. Department of Commerce, National Oceanic and Atmospheric Administration—Coastal Ocean Office.
Marshall, C. M., R. W. Kolb, and J. A. Overdahl. 2009. Monte Carlo techniques in pricing and using derivatives. In *Financial derivatives*. New York: John Wiley & Sons, Inc. Pp. 423–440.
Meltzer, D. 1997. Accounting for future costs in medical cost-effectiveness analysis. *Journal of Health Economics* 16(1):33–64.
Morgenstern, R. 1997. *Economic analyses at EPA: Assessing regulatory impact*. Washington, DC: Resources for the Future.
———. 2011. Reflections on the conduct and use of regulatory impact analysis at the U.S. Environmental Protection Agency. Discussion paper prepared for an RFF Conference, April 7. http://www.rff.org/RFF/Documents/RFF-DP-11-17.pdf (accessed November 21, 2012).
Moss, R., and S. Schneider. 2000. Uncertainties in the IPCC TAR: Recommendations to lead authors for more consistent assessment and reporting. In *Guidance papers on the cross cutting issues of the Third Assessment Report of the IPCC*, edited by R. Pachauri, T. Taniguchi, and K. Tanaka. Geneva: World Meteorological Organization. Pp. 33–51.
NAPEE (National Action Plan for Energy Efficiency). 2008. *National action plan for energy efficiency vision for 2025: A framework for change*. Washington, DC: Environmental Protection Agency, Department of Energy.
Newell, R. G., and W. A. Pizer. 2003. Discounting the distant future: How much do uncertain rates increase valuations? *Journal of Environmental Economics and Management* 46(1):52–71.

———. 2004. Uncertain discount rates in climate policy analysis. *Energy Policy* 32(4):519–529.
NRC (National Research Council). 2002a. Estimating the public health benefits of proposed air pollution regulations. Washington, DC: The National Academies Press.
———. 2002b. *Estimating the public health benefits of proposed air pollution regulations.* Washington, DC: The National Academies Press.
———. 2004. *Valuing ecosystem services: Toward better environmental decision-making.* Washington, DC: The National Academies Press.
———. 2008. *Public participation in environmental assessment and decision making.* Washington, DC: The National Academies Press.
OMB (Office of Management and Budget). 2003. *Regulatory analysis: Circular A–4 to the heads of executive agencies and establishments.* http://www.whitehouse.gov/omb/circulars/a004/a-4.pdf (accessed November 21, 2012).
———. 2005. *Validating regulatory analysis: 2005 report to Congress on the costs and benefits of federal regulations and unfunded mandates on state, local, and tribal entities.* Washington, DC: Office of Management and Budget.
Pavitt, K. 1984. Sectoral patterns of technical change: Towards a taxonomy and a theory. *Research Policy* 13(6):343–373.
Pearce, D., G. Atkinson, and S. Mourato. 2006. *Cost–benefit analysis and the environment: Recent developments.* Paris: OECD Publishing.
Portney, P. R. 1994. The contingent valuation debate: Why economists should care. *Journal of Economic Perspectives* 8(4):3–17.
RTI International. 2008. *EMPAX–CGE model documentation interim report.* Research Triangle Park, NC: EPA.
Santos, S., and S. Edwards. 1990. Comparative study of risk assessment and risk communication practices in western Europe: A report prepared for the German Marshall Fund of the United States.
Shimshack, J. P., and M. B. Ward. 2005. Regulator reputation, enforcement, and environmental compliance. *Journal of Environmental Economics and Management* 50(3):519–540.
———. 2008. Enforcement and over-compliance. *Journal of Environmental Economics and Management* 55(1):90–105.
Sloan, F. A., and C.-R. Hsieh. 2012. *Health economics.* Cambridge, MA: The MIT Press.
Slovic, P. 1987. Perception of risk. *Science* 236(4799):280–285.
Thompson, K. M., D. E. Burmaster, and E. A. C. Crouch. 1992. Monte Carlo techniques for quantitative uncertainty analysis in public health risk assessments. *Risk Analysis* 12(1):53–63.
Viscusi, W., and J. Aldy. 2003. The value of a statistical life: A critical review of market estimates throughout the world. *Journal of Risk and Uncertainty* 27(1):5–76.
Wallmo, K., and D. K. Lew. 2011. Valuing improvements to threatened and endangered marine species: An application of stated preference choice experiments. *Journal of Environmental Management* 9(27):1793–1801.
Williams, M. 1994. Combining revealed and stated preference methods for valuing environmental amenities. *Journal of Environmental Economics and Management* 26:271–292.
Yao, R., and P. Kaval. 2011. Non-market valuation in New Zealand: 1974 to 2005. *Annals of Leisure Research* 14(1):60–83.
Yeager, D. S., S. B. Larson, J. A. Krosnick, and T. Tompson. 2011. Measuring Americans' issue priorities: A new version of the most important problem question reveals more concern about global warming and the environment. *Public Opinion Quarterly* 75(1):125–138.

4

Uncertainty and Decision Making: Lessons from Other Public Health Contexts

The U.S. Environmental Protection Agency (EPA) is not the only agency or organization that must make decisions in the face of uncertainty. Other agencies do as well, and, as is the case with EPA, when making a decision those other agencies must consider the likelihood and magnitude of a risk, the number of people at risk, whether some people are more at risk than others, the likelihood that a given intervention will mitigate the risk, the cost of potential interventions, and the potential consequences of inaction.

A number of decisions about public health interventions that are now well understood were made at a point in time when there were more uncertainties. For example, it is now well accepted that the pasteurization of dairy products eliminates the risk of infections caused by *Campylobacter jejuni*, *Salmonella* species, and other pathogens (FDA, 2011a); that fortification of foods with vitamins and minerals decreases the health consequences of vitamin and mineral deficiencies, e.g., that the fortification of wheat products with folate decreases neural tube defects (Darnton-Hill and Nalubola, 2002); that vaccination against common childhood infections prevents serious morbidity and mortality (Bonanni, 1999); and that prenatal screening for HIV infection facilitates the immediate postdelivery administration of antiviral agents to prevent HIV infection in an infected baby (Anderson and Sansom, 2006). However, not all of those interventions were unanimously accepted when first proposed or implemented, primarily due to uncertainties surrounding the possible benefits, risks, costs, feasibility, and public values. Many of those uncertainties have been reduced through research, including research on the effects of the interventions or treatments that were implemented. In contrast, other interventions that were once thought beneficial, such as bed rest after childbirth or a heart attack, were found not to be beneficial once uncertainties were reduced.

In this chapter the committee reviews the decision-making tools and techniques from a number of different areas of public health, focusing on how uncertainty is taken into account in decisions. In particular, these reviews are in response to two of the questions in the committee's charge: "What are promising tools and techniques from other areas of decision making on public health policy? What are benefits and drawbacks to these approaches for decision makers at EPA and their partners?" The committee could not review all organizations that make public health decisions or all decision-making processes, so it focused on selected agencies and organizations that, as does EPA, assess benefits and risks to human health (and in some cases technological, economic, and other factors), identify uncertainties, and make regulatory or policy decisions on the basis of those analyses. The chapter begins with a general discussion of the decision-making processes at a number of government agencies and organizations. It then uses case studies to illustrate how different agencies and organizations have made difficult regulatory or policy decisions while accounting for uncertainties.

UNCERTAINTY AND PUBLIC HEALTH DECISIONS

A number of U.S. agencies play important public health roles that involve weighing evidence and taking into account uncertainties in the making of a policy or regulatory decision that affects public health. Table 4-1 summarizes the processes and methods used by different public health agencies and organizations to evaluate the human health risks and benefits and other factors influencing the decisions, along with their inherent uncertainties. As can be seen in the table, many organizations have no formal guidance materials related to their decision-making processes, and many do not conduct formal uncertainty analyses.

Within the U.S. Food and Drug Administration (FDA), some divisions—such as the center responsible for overseeing drug approvals and post-marketing safety, the Center for Drug Evaluation and Review, and the center responsible for overseeing medical devices, the Center for Devices and Radiological Health—have published guidance material on risk assessments. Historically, however, neither center provides a thorough discussion of uncertainty analyses or of the communication of those uncertainties along with FDA decisions, although a recent report for FDA has highlighted the importance of communicating the uncertainties in the agency's decisions and the data that underlie them (Fischhoff et al., 2011). The Occupational Safety and Health Administration (OSHA) and the Nuclear Regulatory

TABLE 4-1 Assessment of Risks, Benefits, Other Decision-Making Factors, and Uncertainty at Selected Public Health Agencies and Organizations

Agency/ Organization	Method of Assessing Risks	Uncertainty Analyses
EPA[a]	Conducts quantitative assessments of risks. The assessment method varies depending on the nature of the exposure (for example, inhalation exposure vs. ingestion) and the endpoint of concern (for example, cancer vs. non-cancer endpoints). Has published extensive number of detailed guidance documents and other materials related to its assessment methods and assessments of risks, benefits, and other factors related to individual agents and regulatory decisions. Participated in interagency working group on risk assessment guidelines.	Conducts extensive quantitative uncertainty analyses of the risks of individual chemical or other agents. The uncertainty analyses of human health risk estimates often includes the uncertainties in the • dose–response assessment, • exposure assessment, • toxicity assessment, and • risk characterization. Has conducted some assessments, including analysis of the uncertainty in estimates of benefits and costs.
FDA–CFSAN	Conducts quantitative assessments of risks, including product-specific assessments, pathogen- and chemical-specific assessments, product-pathway assessments, and risk-ranking assessments (for example, *Listeria monocytogenes* in ready-to-eat foods, and methylmercury in seafood). Published guidance for risk assessments for food terrorism (FDA, 2012a).	Uncertainty analyses vary among the assessments, with some having qualitative and some having quantitative assessments. Some analyses have estimated the effects of different regulatory actions (for example, *Listeria monocytogenes* assessment in FDA, 2003). Food terrorism and vulnerability assessment guidance discusses the fact that uncertainty exists, but does not provide formal guidance for analysis of uncertainty.

continued

TABLE 4-1 Continued

Agency/ Organization	Method of Assessing Risks	Uncertainty Analyses
FDA–CDER	Has published guidance for industry on premarketing risk assessment (for example, FDA, 2005).	Discusses the fact that uncertainty exists, but does not present any formal guidance for analysis of uncertainty.
	Has published guidance on risk communication with the public in the context of drug safety (FDA, 2012b).	The guidance does not contain a specific discussion of the communication of uncertainty.
FDA–CDRH	Has published guidance for industry for benefit–risk determinations (FDA, 2012c).	Guidance discusses the sources of uncertainty in the science supporting estimates of human health risks and benefits. There is no guidance related to how to analyze uncertainties.
CDC–ACIP	Uses the GRADE system to review and classify evidence (Ahmed et al., 2011).	The GRADE system includes a discussion of the strengths and limitations of the evidence. Depending on the information available, detailed uncertainty information, including uncertainty in the analysis of costs and benefits, is considered in ACIP's recommendations.
AHRQ–Evidence Based Practice Centers	Categorizes the strength of the evidence related to medical interventions using a process based on the GRADE system.	The categories used include a qualitative discussion of the uncertainties in evidence.
OSHA	Conducts quantitative assessments of risk estimates for different exposures (exposures might be, for example, individual chemical exposures, noise exposure, or job descriptions).	Some assessments include some quantitative analyses of uncertainties (such as the presentation of upper and lower bounds on estimates or the evaluation of the effect of using different models to generate estimates).
FSIS[a]	Conducts qualitative, semi-quantitative, and quantitative assessments of human health risks (for example, the *Listeria monocytogenes* risk assessment in FDA, 2003).	The uncertainty analysis varies among the assessments; some include qualitative analyses and some include quantitative analyses, sometimes including analyses of the effects of different regulatory actions (for example, the *Listeria monocytogenes* risk assessment in FDA, 2003).

TABLE 4-1 Continued

Agency/Organization	Method of Assessing Risks	Uncertainty Analyses
Nuclear Regulatory Commission	Conducts quantitative, probabilistic risk assessments to estimate the likelihood and consequences of different events to help develop "risk-informed, performance-based regulations" (NRC, 2012).	Conducts uncertainty and sensitivity analyses in its assessments.
WHO–IARC	Publishes IARC Monographs, which evaluate the increased risk of cancer associated with environmental factors (including chemicals, complex mixtures, occupational exposures, physical agents, biological agents, and lifestyle factors). The monographs include a qualitative assessment to classify environmental factors into groups on the basis of the evidence of carcinogenicity.	Monographs include a qualitative discussion of the uncertainties in the evidence and identify data gaps.
WHO–FAO	Has published detailed guidance and assessments for microbial risk characterization in food that include qualitative, semiquantitative, and quantitative human health risk assessments. Has published quantitative assessments of the health risks associated with various chemicals in food (WHO, 2013).	Detailed discussions of the uncertainties in estimates of health risks and analyses of those uncertainties. Some assessments discuss economic factors in decisions, including uncertainties in economic analyses. Some assessments include discussion of risk communication. Uncertainties are discussed, but no quantitative assessments of uncertainties.

NOTES: ACIP, Advisory Committee on Immunization Priorities; AHRQ, Agency for Healthcare Research and Quality; CDC, Centers for Disease Control and Prevention; CDER, Center for Drug Evaluation and Review; CDRH, Center for Devices and Radiological Health; CFSAN, Center for Food Safety and Nutrition; EPA, Environmental Protection Agency; FAO, Food and Agriculture Organization of the United Nations; FDA, Food and Drug Administration; FSIS, Food Safety and Inspection Service; GRADE, Grading of Recommendations Assessment, Development and Evaluation; IARC, International Agency for Research on Cancer; OSHA, Occupational Safety and Health Administration; WHO, World Health Organization.

[a]EPA and FSIS, in conjunction with other public partners published microbial risk assessment guidelines (USDA/FSIS and EPA, 2012). The guidelines discuss uncertainty, uncertainty analysis, and how to communicate uncertainty for risk characterization.

Commission discuss and use uncertainty analyses when formulating regulations, as does the Agency for Toxic Substances and Disease Registry of the U.S. Centers for Disease Control and Prevention (CDC) when it conducts its health assessments. At the international level, the International Agency for Research on Cancer of the World Health Organization (WHO) evaluates the evidence for the carcinogenicity of different agents and classifies those agents into different categories according to their estimated carcinogenicity; uncertainties are presented qualitatively when discussing the gaps in evidence.

Both the Food Safety and Inspection Service (FSIS) and the Center for Food Safety and Nutrition (CFSAN) at FDA have published—individually and jointly—a number of assessments of the health risks associated with chemical or biological agents in different foods. Those assessments often contain quantitative analyses of uncertainties and sensitivity analyses. In addition, an interagency working group[1] has published draft guidelines for microbial risk assessments for food and water. The guidelines discuss the analysis and communication of uncertainties in risk assessments. WHO and the Food and Agriculture Organization (FAO) of the United Nations have also published guidance for the characterization of the risks from microbial contamination of food. That guidance discusses qualitative and quantitative human health risk assessments and the analyses of uncertainties in those assessments. They also discuss economic analyses to support decision making and the concomitant uncertainties in those analyses. Other uncertainties are not discussed, nor are issues related to the communication of uncertainties in the assessments of health risks and economics.

The U.S. Preventive Services Task Force (USPSTF), an independent task force that is supported and administered by the Agency for Healthcare Research and Quality, uses an evidence-based approach to evaluate health care interventions and make recommendations for clinical practices, including medical screening tests. To do so, the USPSTF uses an adapted version of the Grading of Recommendations Assessment, Development and Evaluation (GRADE) system that qualitatively characterizes as high, moderate, or low the likelihood that a practice or treatment is beneficial. A working group designed the GRADE system as "a common, sensible and transparent approach to grading quality of evidence and strength of recommendations" (GRADE Working Group, 2012b). A number of different organizations, such as the USPSTF and CDC's Advisory Committee on Immunization Practices, have used the GRADE system to characterize evidence and recommendations (GRADE Working Group, 2012a). The use of GRADE when making recommendations related to vaccines is briefly discussed later in this chapter.

---

[1] The working group included representatives from FDA, FSIS, and EPA.

The organizations that use more sophisticated uncertainty analysis, such as the Nuclear Regulatory Commission, FSIS, CFSAN, OSHA, and FAO, do so using methods and approaches that are similar to those used by EPA. A few organizations discuss the presence of uncertainty in economic analyses, but even those organizations do not explicitly discuss how or whether that uncertainty affected their decisions. Furthermore, they rarely consider factors other than health risks, health benefits, and economic analyses in their decision-making process. Many of the organizations elicit input from stakeholders through public meetings and comments on proposed action, much as EPA does; they do not, however, set forth an explicit process for incorporating uncertainties, such as a heterogeneity of stakeholder perspectives, into decision making.

In reviewing the processes of these public health agencies and organizations, the committee identified a number of assessments or decisions that illustrate the techniques and approaches that have helped—or, in one instance, handicapped—decision makers in their efforts to make decisions in the face of uncertainty. These cases include the following, which are discussed below: (1) the assessment of the health effects associated with secondhand smoke; (2) FSIS and FDA's assessment of regulations related to *Listeria monocytogenes*; (3) FSIS's assessment of the human health risks associated with bovine spongiform encephalopathy; (4) FSIS's and FDA's decisions surrounding the contamination of the food supply with melamine; (5) FDA's decisions related to the diabetes medication Avandia® (rosiglitazone); and (6) assessments related to vaccinations. The committee did not attempt to develop a thorough evaluation or critique of each case; rather, it focused on aspects of the different cases that demonstrate useful approaches to evaluating and considering uncertainty in regulatory or policy decisions.[2]

SECONDHAND SMOKE

Smoking bans that limit exposures to secondhand smoke (SHS) have been enacted in many places despite some stakeholders pointing to uncertainties in economic and other data as well as to uncertainties in estimates of health risks as a reason not to enact bans. Those uncertainties are thought to have been generated or at least exaggerated by the tobacco industry (Muggli et al., 2003; Ong and Glantz, 2000; Tong and Glantz, 2007). This section discusses what evidence was available on the economic

---

[2] The committee uses these cases to illustrate the types of analyses and processes conducted in public health settings that can facilitate decision making. The committee is not endorsing, commenting on, or drawing any conclusions about the appropriateness or correctness of the regulatory or policy decisions themselves.

impacts and public acceptance of smoking bans at the time of decisions and what lessons can be learned from the implementation of these bans.

Human Health Risks

Although many of the human health risks associated with cigarette smoking were well established by the 1960s (U.S. Surgeon General's Advisory Committee on Smoking and Health, 1964), it was not until 1986 that a surgeon general's report concluded that SHS increased the risk for many different adverse health outcomes (HHS, 1986). The evidence of the risks from SHS comes from environmental chemistry and toxicology, including animal models of disease, as well from as observational studies (most of which were case-control studies or meta-analyses of those case-control studies). Federal and state human health risk assessments have concluded that SHS is harmful to humans (Cal EPA, 2005; EPA, 1992; NTP, 2011). Most of the risk-assessment findings were based on quantitative, well-conducted studies, although the findings were not always consistent among the studies (see HHS et al. [2006] and IOM [2010a] for reviews of the studies). Concerns about variations in findings for a specific condition were allayed by the large number of studies, their general consistency, and the results of a number of meta-analyses conducted.

Despite the scientific evidence indicating adverse effects of SHS—including EPA's assessment of environmental tobacco smoke (EPA, 1992), some individuals and groups, many of whom had financial interests in not having smoking bans, called the evidence into question (Oreskes and Conway, 2010). Similarly, public comment periods on health risk assessments and proposed policies and regulations were often dominated by individuals or groups criticizing the studies who were often allied with the tobacco industry, and tobacco industry documents indicate that they had a strategy of maintaining the scientific debate around the health effects of secondhand smoke (Bryan-Jones and Bero, 2003). To set smoking policies, therefore, decision makers had to distinguish between true uncertainties in the evidence and unfounded criticisms of the evidence motivated by financial interests, and they had to not only consider the results of each study, but also carefully scrutinize the quality of each study under consideration.

Economic Factors

One economic factor that was taken into account when considering smoking bans was the potential economic effects on the establishments that would be subject to the bans (for example, bars and restaurants). Before the advent of state and local regulation, few studies had evaluated the economic effects that smoking restrictions and bans might have on those establishments. Because detailed studies of the economic consequences of likely regulations and policies were often unavailable, there was also no characterization of the uncertainties surrounding those economic factors. Legislatures were left to make decisions on environmental controls for SHS exposure in the face of large uncertainty and intense lobbying. As smoking bans were enacted and implemented, studies have looked at the economic consequences of the bans, for example, on restaurants and bars (Glantz and Charlesworth, 1999), which has decreased the uncertainty around the economic factors.

Public Acceptance

In accordance with national or local rules, decisions on smoking restrictions and bans generally included the opportunity for the public to comment on the proposed policies and on the science underlying them (Bero et al., 2001). As mentioned above, many of the people commenting spoke out against the policies, but some of that opposition was orchestrated by the tobacco industry and allied parties (Mangurian and Bero, 2000). In addition, several commentators on the process suggested that there was "burnout" by the public on SHS issues and a loss of advocacy that came with many years of direct cigarette regulation, particularly at the local level (WHO, 2006). Furthermore, the national environmental and public health organizations and agencies that could have supported local and state regulations often did not weigh in strongly, possibly because of a coordination (Bero et al., 2001). Those aspects increased the uncertainty about the percentage of people and which sectors of the public were for or against smoking bans and restrictions.

Further problems with the interactions with stakeholders may have been caused by communication issues, including a lack of communication about the uncertainties surrounding the issue. Most communication with stakeholders about uncertainty used standard statistical presentations of epidemiological studies and meta-analyses (Hackshaw et al., 1997; Law et al., 1997), and there appears to have been little attempt in the federal (EPA, 1992; NTP, 2011) and state risk analyses (see, for example, Cal EPA [2005]) to present uncertainty in lay terms. The dose–response phenomenon was also not discussed extensively, with the exception of questions concerning the relevance of studies of home exposures to social exposures. Those discussions could have led the public to believe that the extent and implications of uncertainties in the data and analyses were greater than they actually were.

Decisions in the Face of Uncertainty: Lessons Learned

Given the lack of any known health benefits from exposure to SHS, health assessments centered on the risks associated with exposure to SHS. Despite a large amount of evidence related to those risks, discussions often focused on the uncertainties in the evidence rather than on its consistencies; individuals and groups with a financial stake in blocking smoking restrictions and bans often drove those discussions (Bryan-Jones and Bero, 2003). There was large uncertainty about the potential costs from lost revenues to establishments subject to bans and about the financial benefits from avoided medical costs. The discussions of smoking bans also raised social issues related to infringing on personal, voluntary behaviors and personal rights, with a large amount of heterogeneity in people's opinions on those issues.

States and local jurisdictions where there was either the political will or higher public acceptance of bans were the first to implement smoking restrictions and bans. Researchers took advantage of some of those bans to investigate whether they were associated with any health effects, to study the public reaction to the bans, and to see whether the bans had any economic consequences on establishments covered by the bans. Epidemiology studies indicated that smoking bans or restrictions were associated with decreases in adverse cardiovascular events (Barone-Adesi et al., 2006; Bartecchi et al., 2006; Cesaroni et al., 2008; IOM, 2010a; Juster et al., 2007; Khuder et al., 2007; Lemstra et al., 2008; Pell et al., 2008; Sargent et al., 2004; Seo and Torabi, 2007; Vasselli et al., 2008). Research surveys showed that the public approval of various state and local laws and regulations was generally, although not uniformly, positive after implementation, both in the United States and other countries (Borland et al., 2006; Kelly et al., 2009; Miller et al., 2002; Pursell et al., 2007; Tang et al., 2003). For example, in 2000 73.2 percent of people surveyed in California who had visited a bar at least once in the previous year approved of California's smoke-free laws, up from 59.8 percent in 1998, the year that a ban of smoking in all bars was implemented (odds ratio [OR] = 1.95; 95 percent confidence interval [CI] = 1.58, 2.40) (Tang et al., 2003). Studies of the economic effects of the bans on restaurants and other establishments decreased the economic uncertainties related to smoking restriction and bans (Glantz and Smith, 1994, 1997; Hyland et al., 1999; Sciacca and Ratliff, 1998; Scollo et al., 2003). With the decreased uncertainty provided by all these types of studies, other state and local regulators had stronger evidence on which to base their decisions. As of October 5, 2012, more than 3,581 municipalities had laws that restrict where smoking is allowed, and "36 states, along with American Samoa, the Northern Mariana Islands, Puerto Rico, the U.S. Virgin Islands, and the District of Columbia" had workplace laws that restrict smoking (ANRF, 2012). U.S. efforts to characterize the risks of environmental tobacco smoke have also been used to support international smoking bans.

The passing and implementation of smoking bans provides many lessons for EPA and for other regulators. First, it emphasizes the importance of scientists and policy makers scrutinizing the quality of individual studies as part of appropriately determining the overall weight of the evidence and the uncertainty in it. Second, it demonstrates the need to consider the sources of scientific criticisms and uncertainties that are raised and to separate valid scientific criticisms from invalid ones. Third, it emphasizes that when considering economic factors and other factors, such as public acceptance, uncertainty based on anecdotal concerns about potential financial consequences might not reflect the actual effects of a regulation. Fourth, it illustrates the heterogeneity in public values and how acceptance of health-protective policies can shift over time, leading to new societal norms.

LISTERIA MONOCYTOGENES

Listeria monocytogenes is a bacterium that causes listeriosis, a potentially fatal bacterial infection that can result from eating food contaminated with the bacterium (FDA and FSIS, 2003a). Listeriosis primarily affects pregnant women, older adults, and persons with weakened immune systems (FDA and FISIS, 2003a). Infections during pregnancy can lead to premature delivery, infection of the newborn, or stillbirth. Death occurs in 20 percent of cases of listeriosis (Swaminathan and Gerner-Smidt, 2007); CDC estimates that *L. monocytogenes* causes nearly 1,600 illnesses each year in the United States, including more than 1,400 hospitalizations and 255 deaths (Scallan et al., 2011). FDA and FSIS collaborated, in consultation with the CDC, to conduct a risk assessment of *L. monocytogenes*. In this section, the committee discusses that risk assessment and the uncertainty analyses in it and also discusses how FDA has used the results of that risk assessment to refine its policies around the control of *L. monocytogenes* in different food products within its regulatory purview.

Regulatory Background

L. monocytogenes can contaminate food contact surfaces and also non-food contact surfaces, such as floors and drains in food-processing facilities. The growth of *L. monocytogenes* is more difficult to control than the growth of most other bacteria. Temperatures at or below 40°F control the growth of most bacteria, but *L. monocytogenes* survives on cold surfaces and can multiply slowly at 32°F; temperatures of 0°F are required to completely stop *L. monocytogenes* from multiplying (FDA and FSIS, 2003b).

Both FDA and FSIS have regulations related to *L. monocytogenes* in ready-to-eat (RTE) foods. FSIS has a zero-tolerance policy for *L. monocytogenes* on RTE meat products within its regulatory purview, such as hot dogs and luncheon meats (FSIS, 2003). Until FDA published proposed draft guidelines in 2008 that established two categories of RTE foods, it had a zero-tolerance policy for *L. monocytogenes* in all RTE foods within its regulatory purview. Under those zero-tolerance policies, the presence of *L. monocytogenes* indicated the product was "adulterated" and unfit for commerce (FDA, 2008a).

The food industry argued that not all RTE foods support the growth of *L. monocytogenes* equally and, therefore, that not all RTE foods should be subject to the same regulations. The food industry identified low pH, low water activity, and the presence of an "inhibitory" substance as factors that inhibit the growth of *L. monocytogenes*. The industry also argued that the risk of listeriosis depends on the type and frequency of RTE consumption as well as on home refrigeration factors (i.e., temperature and duration). In essence, the industry was arguing that the variability in food susceptibilities and the uncertainty in home refrigeration had not been considered in the regulations.

In light of those arguments, FDA reviewed its zero-tolerance policy for *L. monocytogenes*. The agency was faced with determining whether or not to relax the zero-tolerance standard for some foods and, if they were relaxed, what guidance it should issue to industry for controlling *L. monocytogenes* (FDA, 2008a).

Human Health Risks

The 2003 FDA/FSIS risk assessment was designed to predict the relative risk of listeriosis from eating certain ready-to-eat foods among people in three age-based groups: perinatal (16 weeks after fertilization to 30 days after birth), elderly (60 years of age and older), and intermediate-age (general population, less than 60 years of age) (FDA and FSIS, 2003a,b). The assessment evaluated 23 categories of foods considered to be the principal potential sources of *L. monocytogenes*. In particular, it evaluated the heterogeneity among different foods and different age groups of people. Consistent with previous assessments, the 2003 assessment concluded that foodborne listeriosis is a moderately rare but severe disease, and it supported the findings from epidemiologic investigations of sporadic illness and outbreaks of listeriosis that certain foods (for example, pâté, fresh soft cheeses, smoked seafood, frankfurters, and foods typically purchased from deli counters) are potential vehicles of listeriosis for susceptible populations.

The assessment estimated the human health risks for different exposures. It also used various scenarios—different food consumption rates, different growth rates, different contamination rates, and so on—to evaluate which points in the farm-to-table continuum were most susceptible to contamination or had the greatest potential for risk mitigation. Through those sensitivity analyses, the assessment identified five main factors that affect consumer exposure to *L. monocytogenes* at the time of food consumption: (1) the amount and frequency of consumption of a food, (2) the frequency and levels of *L. monocytogenes* in RTE food, (3) the potential to support growth of *L. monocytogenes* in food during refrigerated storage, (4) refrigerated storage temperature, and (5) the duration of refrigerated storage before consumption. Those factors point toward several control strategies to mitigate the risks of listeriosis, which ranged from "reformulation of products to reduce their ability to support the growth of *L. monocytogenes*" to "encouraging consumers to keep refrigerator temperatures at or below 40°F and reduce refrigerated storage times" (FDA and FSIS, 2003a, p. 27).

Rather than providing a single risk estimate, the health risk assessment provided a range of estimates using sensitivity analyses and probabilistic methods for the different food categories, the populations with different susceptibilities to listeriosis, and the strains of *L. monocytogenes* with varying virulence. The assessment "attempt[ed] to capture both the variability inherent in the incidence of foodborne listeriosis and . . . the uncertainty associated with the data analysis" (FDA and FSIS, 2003a, p. 15). Presenting the different sensitivity analyses allowed decision makers to target strategies to mitigate risks for different populations and food categories. For example, specific strategies could be developed to prevent exposures in pregnant women, the elderly, and susceptible individuals within the intermediate-age group. In addition to the uncertainty analyses discussed above, FDA/FSIS discussed other uncertainties that remained, including the need for evidence related to changes in food processing, distribution patterns, preparation, and consumption practices.

FDA used the results of the risk assessment and its analyses of variability to develop regulations that differentiated between foods that pose higher and lower risks for listeriosis. In February 2008, FDA issued a draft Compliance Policy Guide (FDA, 2008a) that proposed two risk-based limits for *L. monocytogenes* in RTE foods, differentiating between foods that support the growth of the pathogen and those that do not. That regulation takes into account the conclusions from the sensitivity analyses in the assessment that the risks from foods with a pH less than or equal to 4.4, foods with water activity less than or equal to 0.92, and frozen foods do not support the growth of *L. monocytogenes* and, therefore, pose very low risk of listeriosis.

Stakeholder Input

The process for developing the FDA/FSIS risk assessment included a period for public comment on a January 2001 draft of the *L. monocytogenes* risk assessment (FDA and FSIS, 2003b). The final risk assessment summarized the changes made to the draft in response to the comments received. Many of those changes reflected decreased uncertainties as a result of the feedback received. They included changes to the food categories to better incorporate characteristics that contribute to the support of growth of *L. monocytogenes* (for example, moisture content and pH), updated data on contamination, growth rates of *L. monocytogenes* in different foods and for different storage durations, the frequency and prevalence of *L. monocytogenes* on different foods, consumer habits, and modifications to the model used.

Decisions in the Face of Uncertainty: Lessons Learned

The assessment of *L. monocytogenes* and FDA's use of that assessment highlight how analyses of health risks that account for uncertainties—such as those for different food types, different storage conditions, and different susceptible populations—can provide decision makers with information to help design policies that target mitigation strategies to the greatest risks, either for specific foods with a higher likelihood of being associated with human illness, or for populations that are more susceptible to illness. The detailed and specific risk characterization allowed FDA and FSIS to develop specific guidance for different foods and to develop outreach strategies to protect the populations at highest risk from consumption of foods contaminated with *L. monocytogenes*.[3] In other words, the assessment and how it was used in FDA's decision highlights the importance of analyzing the heterogeneity in an assessment and demonstrates how—when uncertainty about that heterogeneity is reduced—an agency can better tailor risk-mitigation options.

In light of the complexity of the risk assessment, the agency also evaluated methods for grouping the results for communication purposes (FDA and FSIS, 2003b). The assessment concluded, "One approach that appears to be very useful for risk management/communication purposes is the evaluation of the relative risk ranking results using cluster analysis"

---

[3] A recent outbreak of *L. monoctyogenes* in cantaloupes, however, shows that even lower-risk foods are not immune to contamination. See CDC (2011a) for a description of the outbreak. FDA investigated that outbreak, identified a number of factors that contributed to the outbreak (FDA, 2011c), and highlighted the need for all processors to employ good agricultural and management practices (FDA, 2011c), as are provided in FDA and USDA guidance to industry (FDA, 2008c).

(p. 228). That analysis allowed the development of a matrix to depict five overall risk designations: very high, high, moderate, low, and very low. For example, deli meats are considered very high risk because they were in the high cluster for both per-serving and per-annum consumption.

BOVINE SPONGIFORM ENCEPHALOPATHY

Bovine spongiform encephalopathy (BSE), commonly referred to as mad cow disease, is a chronic degenerative disease that affects the central nervous system of cattle. BSE is one of a number of transmissible spongiform encephalopathies that are caused by infectious agents associated with an abnormally folded protein known as a prion (IOM, 2004). In cattle, the infectious agent is transmitted through ingestion of contaminated feed. In humans, exposure to beef products that are infected with BSE can lead to Creutzfeldt-Jakob disease (CJD), a fatal neurodegenerative disease (IOM, 2004). Because there are no vaccines against or treatments for BSE or CJD and because it is extremely difficult to destroy the infectious agent, preventing the spread of BSE among cattle and preventing cattle infected with BSE from entering the human food supply are key control mechanisms. The potential human health consequences from CJD are severe, and the costs of an outbreak are high. For example, hundreds of thousands of infected animals had to be destroyed, and trade restrictions were instituted by other countries, following an outbreak of BSE in Britain in 1985 (IOM, 2004).

In 1998 the U.S. Department of Agriculture (USDA) requested that the Harvard Center for Risk Analysis (HCRA) evaluate measures to control the spread of BSE among animals and from animals to humans. In response, in 2003 the HCRA developed a model to simulate the consequences of introducing BSE into the United States by various means (Cohen et al., 2003a,b). USDA and the public provided comments on the assessment, and HCRA published an updated assessment in 2005 (Cohen and Gray, 2005), and the results of additional simulations were published in 2006 (Cohen, 2006). The assessment demonstrates how, when faced with uncertainty about the best regulatory option, the effects of different management options can be modeled to inform the decision-making process.

Assessment of BSE Infection Risks

To assess the risks to cattle and to humans from the introduction of BSE in the United States, HCRA designed a model that could predict "the number of newly infected animals that would result from introduction of BSE, the time course of the disease following its introduction, and the potential for human exposure to infectious tissues" (Cohen et al., 2003b, p. vii). The model also incorporated "key processes and procedures that make the spread of disease more or less likely" (p. vii) to estimate the effectiveness of different control measures in stemming the spread of BSE to cattle and decreasing the likelihood of BSE-infected meat entering the human food supply. Table 4-2 lists the processes and procedures evaluated in the 2006 simulations (see notes section in Table 4-2 below). For example, the model was used to estimate the number of infected cows anticipated 20 years after 500 infected animals were introduced into the United States, using different beef consumption rates or with different detection rates for BSE in antemortem inspections. Using the model, HCRA was able to evaluate different scenarios and look at the effects on the human food supply up to 20 years after the introduction of a given number of BSE-infected cattle into the United States.

The model makes it possible to examine the degree to which different parameters affect the estimated risks. For example, the sensitivity analyses demonstrated that the model is very sensitive to the rate of either accidental or intentional misfeeding of cattle with feed containing ruminant protein but that there is a large uncertainty in the rate at which that misfeeding occurs. Changing the incubation time also affects the model outputs;

TABLE 4-2 Estimates of Infected Cases of BSE in the 20 Years Following Introduction of 500 Infected Animals into the United States

		Percentile				
Label	Mean	5th	25th	50th	75th	95th
Base Case	180	33	98	160	240	400
Sensitivity 1	200	38	110	180	270	440
Sensitivity 2	2,600	1,200	1,900	2,500	3,200	4,400
Sensitivity 3	240	38	130	210	330	530
Sensitivity 4	180	33	97	160	240	400
Sensitivity 5	190	36	100	170	260	420
Sensitivity 6	43	6	13	24	60	130
Sensitivity 7	180	33	97	160	240	400
Sensitivity 8	180	33	97	160	240	400

NOTES: Explanations of the sensitivity analyses:
Sensitivity 1 – Pessimistic MBM/feed production mislabeling and contamination assumptions
Sensitivity 2 – Pessimistic misfeeding assumptions
Sensitivity 3 – Pessimistic render reduction factor assumptions
Sensitivity 4 – Higher assumed beef on bone consumption rates
Sensitivity 5 – Pessimistic antemortem inspection BSE detection rates
Sensitivity 6 – Longer incubation period
Sensitivity 7 – Evaluate the importance of the proportion of cattle showing no clinical signs of diseases that are nonambulatory
Sensitivity 8 – Evaluate the importance of the proportion of clinical animals that are nonambulatory
Abbreviations: BSE = bovine spongiform encephalopathy; MBM = meat and bone meal.
SOURCE: BSE Risk Assessment (Cohen, 2006; FSIS and USDA, 2004).

the number of newly infected cattle decreased from 180 to 43 when the incubation period was lengthened by a factor of two. In contrast, other parameters—such as the rates of mislabeling and contamination, render reduction factors, the consumption rates of beef on bone, and the effectiveness of antemortem inspections—had less influence on model outputs (Cohen and Gray, 2005). The 2006 assessment analyzed other regulatory options, ranging from a ban on slaughter for human consumption of all nonambulatory disabled cattle to prohibiting human consumption of the brain, skull, eyes, trigeminal ganglia, spinal cord, vertebral column, and dorsal root ganglia of cattle 30 months of age or older, in order to evaluate the impacts of these options on risks. That assessment also presented the effects of different regulatory compliance rates on risks, and it presented, in addition to means, the 5th, 25th, 50th, 75th, and 95th percentile values for the outputs (see Table 4-2). That information gave decision makers an indication of how the uncertainty in the data used in the assessment might affect the output from the assessment. Overall, the 2006 simulations indicated that the preventive measures enacted by USDA decrease the potential for human exposures but have little effects on the spread of BSE in the U.S. cattle population.

Economic Factors and Public Sentiment

FSIS implemented a number of rules to protect the public's health in the wake of finding a BSE-positive cow in the United States,[4] and it also conducted a regulatory impact analysis of those rules. The analyses used probabilistic models to estimate the "costs and revenues changes (a partial budget analysis) associated with the final rule compared with the baseline; the net total monetary changes of decreased revenues and increased costs versus increased revenues and decreased costs; and the distribution of those net revenues and costs changes among producers" (FSIS and USDA, 2007, p. 24).

The inputs in the economic model included ranges of values for the variables related to the costs for industry to comply with the rule, such as the number of affected establishments, the number of different types of cattle slaughtered (for example, bulls, cows, or steers), the weight of different cattle, the number of days a week a slaughter facility operates, labor costs, the one-time capital costs for equipment, ongoing material costs of compliance, and disposal costs. The analyses presented minimum, maximum, and most likely values for the compliance costs as well as for associated biological parameters. Using that approach, FSIS estimated probability distributions for compliance costs and for the potential reductions in

---

[4] FSIS described the actions as "emergency actions to protect public health" (FSIS/USDA, 2004, p. 1).

the risks to human health for different risk-mitigation options. The agency then calculated the cost-effectiveness ratio for each of the options and the incremental cost-effectiveness ratios for the options. Although the data tables in an appendix of the final FSIS/USDA report presented a range of values—specifically, the minimum, mean, and maximum costs—the tables in the main body of the report showed only the mean (or most likely) costs.

The report identified a number of data gaps in the analyses. For example, it was not known how much it would cost to redesign facilities to allow for the segregation of animals and animal carcasses, nor was it known how many facilities had already implemented such changes; those costs, therefore, were not included in the analysis. The regulatory impact analysis supported most of the rules FSIS had implemented on an emergency basis, although the analysis did lead to changes being made to some aspects of the rule (FSIS and USDA, 2007).

FSIS also held a number of public meetings to obtain input on the various potential impacts of regulatory actions and assessed how best to communicate the issues concerning BSE to the public, given the large uncertainties in both the data and the analyses.

Decisions in the Face of Uncertainty: Lessons Learned

The analyses that FSIS performed in response to the discovery of a BSE-positive cow in the United States provide an example of a decision-driven risk assessment. The probabilistic models were designed to evaluate specific regulatory options and to identify where within the beef-processing system the agency could have the greatest effect on risks. The benefit and cost-effectiveness analyses provided FSIS with information about the comparative costs of different actions to help decision makers focus resources. The assessment demonstrates how analyses can evolve and be updated in response to new information, stakeholder input, emergency situations, and evolving regulations. The agency's actions demonstrate an adaptive management process used in the face of deep uncertainty about the health risks of BSE and the appropriate mitigation strategies necessary to protect against it. FSIS implemented emergency regulations and modified them, as appropriate, after a full regulatory impact assessment was conducted.

CONTAMINATION OF THE FOOD SUPPLY WITH MELAMINE

Meat, Chicken, Eggs, and Catfish

In 2007 FDA, in conjunction with FSIS, conducted an interim assessment of the risks in the human food supply of melamine, which people can get from eating pork, chicken, fish, and eggs. The concern arose because some animals that are raised to be consumed by humans were inadvertently fed animal feed that may have been adulterated with melamine and its analogues. Those compounds had been associated with kidney failure in pets that had eaten contaminated pet food. Because of these concerns, animals on farms where contaminated feed was distributed were held (either voluntarily or by state quarantine) pending an assessment of the risks to human health (FDA and FSIS, 2007).

When the contamination occurred, there was a great deal of uncertainty surrounding the human health risks from melamine. No information was available on the relative potency of melamine and its analogues. It had been hypothesized that melamine might act synergistically with its analogues, but that hypothesis had not been tested. There had been no studies in humans, and high-dose studies in dogs, rats, and mice had shown hepatic toxicity but not liver failure. Furthermore, the mechanism of those effects was not understood, and it was not known if the adverse effects would occur in humans or at lower doses. In the face of those uncertainties, the agencies needed to estimate the risks to humans from the consumption of meat and fish possibly contaminated with melamine and to quickly decide what, if any, action was needed to protect the public's health.

A 13-week study in rats that had been orally exposed to melamine indicated a no-observed-adverse-effect level (NOAEL; see Chapter 2 for an explanation of NOAELs) for bladder stones of 63 mg/kg body weight/day. Given the lack of data and the need for a quick decision, a quantitative analysis of the uncertainty was not possible, and the agencies used safety factors to evaluate the risks. That NOAEL was divided by a safety factor of 100 (two 10-fold safety factors, or uncertainty factors, to account for inter- and intraspecies sensitivity) in order to come up with a tolerable daily intake (TDI)[5] of 0.63 mg/kg body weight/day. The agencies estimated human exposures to melamine for three different scenarios, including a worst-case scenario, and presented the mean and 90th percentile exposures in each scenario based on a consumption of catfish, chicken, eggs, pork, or a combination of all four (FDA and FSIS, 2007). Given the concentrations of melamine measured in samples of meat collected from animals exposed to the melamine-contaminated feed and estimated human exposures from consumption of meat, the agencies concluded that, even using the worst-case consumption scenario that assumed all solid food was contaminated with melamine, the estimated potential exposures were well below the

---

[5] The TDI is defined as the estimated maximum amount of an agent to which individuals in a population may be exposed daily over their lifetimes without an appreciable health risk with respect to the endpoint from which the NOAEL is calculated (http://www.fda.gov/Food/FoodSafety/FoodContaminantsAdulteration/ChemicalContaminants/Melamine/ucm164658.htm (accessed January 3, 2013).

TDI. Given the estimated risks, FSIS decided that products from animals fed contaminated feed would not be considered adulterated and, therefore, could be made available for slaughter (FSIS, 2007).

Infant Formula

In September 2008 FDA learned that some infant formula from a Chinese manufacturer might contain melamine. Consumption of melamine-contaminated infant formula in China had resulted in a reported 52,857 cases of nephrolithiasis (and, in some instances, renal failure), including about 13,000 hospitalizations and 3 confirmed deaths (FDA, 2008d). The exposure scenario and the sensitivity of the population in the case of infant formula were very different from those in the case of the contamination of meat, fish, and eggs. Some of the specific differences were

1. the contaminated product, infant formula, represented the total caloric intake for most of these infants;
2. the exposure was chronic over a number of months;
3. the population exposed to the products consisted of infants and toddlers whose renal systems had not yet been fully developed; and
4. the human exposure had not been mitigated by the melamine passing through the digestive system of an animal (FDA, 2008d).

Many of the uncertainties discussed above in the case of the contamination of meat and other food products were still in place, and once again FDA had to quickly estimate the risks to humans from melamine—in this case, the risks to infants—and to decide what actions, if any, were necessary to protect public health. Additional uncertainties stemmed from the possibility that premature infants with immature kidney function who had been fed formula as the sole source of nutrition were getting a significantly larger exposure to the melamine than was seen in other cases; they could be ingesting more of the chemical per unit body weight than adults who ate the meat and other products, and they could have been exposed for a longer time period than full-term term infants.

Some studies had been published since the 2007 risk assessment (FDA and FSIS, 2007), including studies related to melamine metabolism and studies on the pathology resulting from exposures in pets (cats and dogs). Data since the 2007 interim risk assessment raised further concerns about an increased toxicity from combined exposure to melamine and its analogue cyanuric acid. Because of those unknowns, FDA concluded in its October 2008 assessment that it could not "establish a level of melamine and its analogues in these products that does not raise public health concerns" (FDA, 2008d). Given the risks, infant formula from China was recalled, and FDA initiated a sampling program to test formula and other products for melamine contamination (FDA, 2008b).

In November 2008, FDA updated its assessment after finding infant formula that had very low concentrations of either melamine or one of its analogues, but not both (FDA, 2008e). That contaminated formula, which was manufactured in the United States, contained concentrations ranging from 0.137 ppm of melamine in one product to 0.247 ppm of cyanuric acid in another. Those concentrations were "up to 10,000 times less than the levels of melamine reported in Chinese-manufactured infant formula" (FDA, 2008e). FDA had to estimate the risks to infants, taking into account the uncertainty. To do so, FDA began with the TDI previously calculated for adult exposures to melamine in meat, chicken, eggs, and catfish.

Given the potential for infants to be more sensitive than adults for the reasons discussed above, FDA applied a 10-fold safety factor to the TDI to get a TDI of 0.063 mg melamine/kg body weight/day to evaluate the additional risks. Assuming a worst-case scenario in which all of an infant's total dietary intake (0.15 kg of powdered infant formula) was contaminated with melamine, 100 percent of the diet would have to be contaminated with 1.26 ppm of melamine for an infant to reach the TDI. FDA concluded, therefore, that the "levels of melamine or one of its analogues alone below 1.0 ppm in infant formula do not raise public health concerns" (FDA, 2008e). In light of that conclusion, FDA did not recall infant formula that contained concentrations of melamine below 1.0 ppm, and it continued its sampling program to test for melamine in the food products that it regulates.

Decisions in the Face of Uncertainty: Lessons Learned

As was the case in the contamination of the food supply with melamine, agencies sometimes have to make regulatory decisions in the face of deep uncertainties about the health risks and—because of the potential for imminent public health consequences—with little or no time to investigate those risks. The responses to contamination of the food supply demonstrate how, even in the absence of probabilistic modeling of uncertainty, a human health risk assessment can provide information for an important regulatory decision. The use of scenarios, including a worst-case scenario as above, can help with such decisions.

AVANDIA®

The regulation of some prescription pharmaceuticals provides an example of decision making in the face of many uncertainties. The case of the diabetes medication Avandia (rosiglitazone) illustrates the uncertainties and scientific disagreements surrounding drugs and how FDA makes decisions under those uncertainties. Private industry, government agencies, health professionals, and the public are important stakeholders for those decisions, and each has different roles, responsibilities, levels of scientific literacy, and values.

Premarketing Benefit and Risk Assessments

Before a pharmaceutical can be marketed, it must undergo a lengthy approval process involving the review of clinical trial data on efficacy, toxicity data, and pharmacokinetic profiles. But the information available, especially related to the safety of the drug, is limited at the time of approval. Those limitations include the facts that the products are generally tested in individuals with single illnesses for relatively short periods of time (compared with the lifetime use of some products) and in many fewer people than will be taking the drug once it goes to market (approval can be obtained with as few as 4,000 participants, and possibly many fewer) (IOM, 2012a).

FDA has published general guidance about the approval process for prescription drugs, but that guidance does not include recommendations for formal, quantitative analysis of uncertainties. For example, FDA has guidelines for the design and conduct of premarketing clinical trials (FDA, 1998), but the manufacturers have broad latitude in designing their studies; the FDA scientists who review the material interpret those guidelines on a case-by-case basis for the specifics of each drug. Ultimately, FDA decision makers evaluate the evidence for the benefits and risks of a drug in the context of the public health need for the drug and decide whether to approve it. FDA sometimes convenes advisory committees of experts to make recommendations regarding the approval of drugs (and regarding other topics as well), and FDA advisory committees might interpret the data or the guidelines differently from the FDA reviewers. FDA often, but not always, follows the recommendations of its advisory committees. The vote of an advisory committee on whether a drug should be approved for marketing is often not unanimous, indicating uncertainty in one or more of the factors involved in drug approval (FDA, 2011b).

Regulatory and Study History of Avandia

FDA's regulatory decisions related to Avandia (rosiglitazone) provide another example of decision making under uncertainty. In 1999 FDA approved Avandia for the treatment of diabetes, but it requested that the drug's sponsor conduct further clinical trials because of concerns about the drug's effects on lipids. In 2007 a meta-analysis of the results of the clinical trials raised concerns about an increased risk of cardiovascular events associated with Avandia (Nissen and Wolski, 2007). Following the publication of that meta-analysis, an FDA advisory committee concluded that the use of Avandia "was associated with a greater risk of myocardial ischemic events than placebo," but because of the limitations of the meta-analysis[6] it recommended label warnings and education rather than a withdrawal of the drug. FDA required a boxed warning on the drug's package and also required a long-term randomized controlled head-to-head clinical trial to evaluate the drug's cardiovascular risks. Observational studies carried out over the next few years indicated an elevated risk of adverse cardiovascular events, and in 2010 FDA once again investigated the drug's benefit–risk profile to decide if any changes in the drug's marketing approval conditions were needed (FDA, 2010a). At that time the agency heard from a number of stakeholders about the importance of having the drug available to them, despite the risks of adverse cardiovascular events. After reviewing the scientific evidence and hearing the views of patients and other stakeholders, in September 2010 FDA stopped its required trial and placed a number of restrictions on access to Avandia, but it allowed the drug to remain on the market in the United States (FDA, 2010c).

In making that decision FDA had to weigh the benefits of a widely used drug that was effective at combating diabetes—a large public health problem—with the risks of adverse cardiovascular outcomes—another large public health problem—when great uncertainty existed. As is typically the case given the nature of the data for evaluating drugs, no formal quantitative uncertainty analysis was conducted. In describing the uncertainty, the director of FDA's Center of Drug Evaluation and Review (CDER) stated that

> there are multiple and conflicting signals of . . . risk. . . . The current cardiovascular safety database for [the drug] does not provide an assurance of safety at the level set out in FDA's guidance for marketed . . . drugs. . . . Many highly experienced clinical trialists and methodologists, both within and external to the FDA, who have examined these data find it hard to arrive at definitive conclusions. . . . This uncertainty about the risk of [the drug] is overwhelmingly the most important reason for the differing opinions on what regulatory action should be taken . . . [and] various members of the Committee had quite disparate opinions on these matters. These differences of opinion stem from varied conclusions about the existing data. . . . Similarly, several [Agency] Offices have different recommendations. The Office of Surveillance and Epidemiology . . . recommends market withdrawal. . . . [The] Office of New Drugs recommends additional

---

[6] Most of the committee members agreed that there was at least a strong signal for increased cardiac ischemic risk, although concerns were raised about the short duration of the trials, the quality of the data, the low number of cardiac events, the lack of cardiac event adjudication, and concerns about the heterogeneity of the study population (FDA, 2007).

warnings on the drug label, without restrictions on marketing. . . . The basis of these recommendations is the uncertainty about the existence of the cardiovascular ischemic safety risk.

Despite the lack of clarity in the data, I believe it is most prudent, given the current uncertainty about the safety risk, to restrict access to the product, and ensure that patients and prescribers are fully informed of the evidence of risk, until and unless more information is obtained. . . . The evidence pointing to a . . . risk with [the drug] is not robust or consistent. . . . Nevertheless, there are multiple signals of concern, from varied sources of data, without reliable evidence that refutes them. Additionally, evidence available to date . . . does not reveal a signal of . . . risk with the other . . . drug available on the US market. . . . Therefore, based on this safety information, it is necessary to restrict access . . . until more substantial evidence of its safety becomes available. (Woodcock, 2010)

FDA qualitatively discussed the scientific uncertainties in an unusually transparent manner not only by publishing on its website the CDER director's final decision (in this case, regarding safety information generated after approval), but also by openly discussing the different opinions of agency scientists at a science advisory meeting and by posting nine scientific review documents on the agency website (FDA, 2010b). As detailed in a recent IOM report (IOM, 2012a) those memoranda highlighted scientific disagreements among agency scientists. For example, agency scientists disagreed about how mechanistic data should affect the consideration of different data from humans, whether missing data points were appropriately handled in studies of Avandia, what endpoints should have been used in studies, and how generalizable data from studies are. Those documents present a detailed analysis of the uncertainties in data and the differing opinions that agency scientists had about the policy options available to the agency decision makers. FDA decision makers, in the absence of quantitative analyses of those uncertainties beyond the presentation of different scientific opinions, made the regulatory decision.

Once FDA approves a drug, other stakeholders have one or more decisions to make. For example, payers decide whether they will pay or reimburse for a given drug; health care providers, in discussion with their patients, decide which drug should be prescribed; and the patient decides whether the side effects of a drug are worth the benefit. Each of these decisions includes uncertainties, some of which can be better understood or reduced through more research, but the decisions often need to be made before that research is available. For each of those choices, in a process that can be either formal or informal, organizations or individuals assess the benefits and the risks, identify uncertainties, and make their decisions.

Decisions in the Face of Uncertainty: Lessons Learned

This example illustrates the sorts of pervasive uncertainties that a decision maker can face and the different ways that government scientists can interpret the same sets of data. (See IOM, 2012a, for more details.) Ultimately, the decision maker must chose from a range of options. The Avandia decision illustrates how an agency can transparently represent those different interpretations and explain what mattered for the final decision. Despite the many uncertainties, a decision was made and the rationale explained. The decision demonstrates how stakeholders' input, such as the need to keep a drug available on the market, can be considered in a regulatory decision. It also highlights the limitations of preclinical data, and the importance of reevaluating and revisiting decisions as new information emerges.

VACCINATION DECISIONS

Decisions about vaccination of the public, such as whom to vaccinate and when, are often associated with uncertainty. Those decisions might have to be made for relatively low-probability, high-consequence events, such as vaccinations to avoid a pandemic, as well as for relatively high-probability events with potentially high consequences, such as vaccinations to prevent infection with the human papillomavirus (HPV),[7] the causative agent for cervical cancer. This section discusses some of the considerations and lessons learned from decisions about influenza and HPV vaccination programs.

Pandemic Prevention and Vaccinations

Decisions related to low-probability, high-consequence events are particularly challenging and fraught with uncertainty. The preparations needed for pandemic influenza or bioterrorist events are examples of such a decision problem. Pandemic influenza occurs when the influenza virus, which circulates around the globe constantly, undergoes a major change in its antigenic presentation (antigenic shift), leaving many in the population with

---

[7] The overall prevalence of HPV was estimated at 42.5 percent in 14- to 59-year-old females in the United States using data from the 2003 through 2006 National Health and Nutrition Examination Surveys (NHANES) (Hariri et al., 2011). The estimated prevalence of HPV types 6, 11, 16, and 18 was 8.8 percent (95 percent confidence interval [CI], 7.8–10.0 percent). Types 16 and 18 are estimated to be responsible for 70 percent of cervical cancers (Dunne et al., 2011).

no immunity.[8] Influenza pandemics have occurred three times in the 20th century and in the first decade of the 21st century. Judging when a pandemic will occur and how to prepare for it is great challenge in the face of tremendous uncertainty. Because of the time period required to develop and administer a vaccine, decisions about how to respond to emerging information occur with great uncertainty about various issues, such as whether the strain identified is circulating widely, whether it is a strain that would incur serious morbidity and mortality, whether a vaccine could be developed in time to mitigate the virus, whether the vaccine would be safe, and whether citizens would get vaccinated.

LESSONS FROM HISTORY

The government's swine flu program that began in March 1976 and ended in March 1977 illustrates some of the problems of making decisions when there is great uncertainty. The decision process that went into the pandemic preparedness and response for the 1976 swine influenza epidemic has widely been viewed as a failure. Neustadt and Fineberg (1978) analyzed that decision-making process with the goal of informing future decisions that might have to be made under similar levels of uncertainty. Some of those lessons are also applicable to EPA.

In 1976 a previously unknown swine flu virus was identified from four cases of influenza, including one fatal case, at a training center for Army recruits (Neustadt and Fineberg, 1978). Concerns were raised because of the possibility of human-to-human transmission; because of a lack of built-up antibodies from previous, similar infections in anyone under 50 years of age; and because of memories of the very virulent strain of swine flu that led to the pandemic of 1918 which killed 500,000 mostly young and able-bodied people in the United States alone. In light of those concerns, a decision was made to implement a mass vaccination plan for all Americans. The pandemic never materialized, vaccine production was delayed, the population did not rush to get vaccinated, and after a rare neurologic side effect—Guillain-Barré syndrome—was identified, the vaccination program was ended.

Uncertainties surrounded a number of issues: the infectivity of the influenza virus, the appropriate dose for children, the production schedules of the vaccine, and the ability to carry out—and the public's willingness to participate in—mass vaccinations. Those uncertainties were considered in the decision-making process and were discussed by policy makers, vaccine

---

[8] Antigenic drift refers to the small changes in antigenic presentation that the virus undergoes constantly. Exposed people have some immunity from previous years' exposures. Antigenic shift is a much more serious change in the influenza virus and can lead to a pandemic.

manufacturers, and scientific, medical, and public health experts; they were not, however, thoroughly analyzed and adequately considered. The broad lessons that Neustadt and Fineberg (1978) gleaned from their evaluation of the immunization program were (1) the need to build in points at which a program will be reviewed, that is, to take an iterative approach to implementation if possible; (2) the need to consider when a plan is feasible; (3) the need to be prepared to deal with questions from the media and the public, but not to let the possibility of questions dictate the decision; (4) the importance of an agency's credibility; and (5) the need to think twice about medical knowledge, including the uncertainties in that knowledge. That last point offers perhaps the most important lesson from the swine flu experience: the importance of framing the decision in terms of what data are available, what uncertainties there are in the data, and, importantly, what new facts or evidence, if at hand, would lead to a different decision and when such different decisions would be made (Neustadt and Fineberg, 1978).

Today's World

Given the many uncertainties about vaccine availability, effectiveness, and utilization, in recent years much of the discussion about recent pandemic preparedness has involved the issue of public values regarding who should be vaccinated in the early days of the pandemic when vaccine availability is limited. Engaging the public to explore their values regarding vaccine distribution has revealed important differences between what scientists thought and what the general public thought about who should be vaccinated first. For example, in a pilot project designed to elicit input from "approximately 300 citizens and stakeholders in different parts of the United States" there was strong agreement among the participants that "'Assuring the Functioning of Society' should be the first goal and 'Reducing Individual Deaths and Hospitalizations Due to Influenza' should be the second priority goal. . . . There was little support for other goals to vaccinate young people first, or to use a lottery system or a first come first served approach as top priorities" (Bernier and Marcuse, 2005, p. 7).

The uncertainty about vaccine availability has also led to intensive efforts to model the effects of community containment (such as school closures, the discouragement of public gatherings, and such hygienic measures as the use of respiratory masks and advice about hand washing) on the spread of the pandemic. The evidence base for the use of those measures has been quite meager (IOM, 2006), and has, in part, depended on analysis of historical data from the 1918–1919 pandemic. Despite the many uncertainties, however, the federal government needed to provide advice to state governments and to the public, so a national plan was developed, and when a new pandemic strain emerged in 2009, the plan was implemented. A recent IOM report (2012b) describes the initial version of a modeling tool to help prioritize the development of vaccines. The modeling tool could be used to identify the likely effects of uncertainty in different factors that play a part in determining the priority of developing a vaccine.

The Human Papillomavirus Vaccine

During the past 20 years researchers have identified HPV as the causative agent in cervical cancer and have characterized the virus and its components (see IOM, 2010b, for an overview), and pharmaceutical companies have developed, tested, and marketed a vaccination against the virus (Baer et al., 2002; Carter et al., 2000; Harper et al., 2004; Harro et al., 2001; Petter et al., 2000). In 2006 the FDA approved the marketing of Gardasil, the first vaccine against HPV, for use in 9- to 26-year-old females. The vaccine and vaccination programs are expensive, however, and CDC's Advisory Committee on Immunization Practices (ACIP) was faced with taking costs into consideration and making a recommendation about who should be vaccinated and when they should be vaccinated.[9]

Researchers have evaluated the costs associated with different vaccination programs in preventing cervical cancer and have outlined the uncertainties in those costs. The annual cost from HPV in the United States was estimated to be $4 billion dollars or more, including the costs associated with the management of genital warts, costs associated with cervical cancer, and costs associated with routine cervical cancer screening and the follow-up of abnormal Pap smears (Markowitz et al., 2007). Models have been used to predict decreases in Pap test abnormalities, cervical cancer precursor lesions, and cervical cancer rates in order to estimate the benefits of vaccination. The ACIP considered four published studies on cost effectiveness (Elbasha et al., 2007; Goldie et al., 2004; Sanders and Taira, 2003; Taira et al., 2004). Markowitz et al. (2007), Sanders and Taira (2003), and Goldie et al. (2004) estimated cost per quality-adjusted life-year (QALY) using Markov models, and Taira et al. (2004) and Elbasha et al. (2007) "applied dynamic transmission models to incorporate the benefits of herd immunity in estimating the cost effectiveness of HPV vaccination" (Markowitz et al., 2007, p. 15). Uncertainties were considered and incorporated in those cost-effectiveness evaluations. Goldie et al. (2004) used a range of inputs for the rates of incidence and the clearance of HPV infection, the natural history of cervical intraepithelial neoplasia (CIN), the natural history of

---

[9] In addition to the estimated health benefit and risk estimates and the economic factors related to the HPV vaccine, there were a number of social and political issues related to whether vaccinations should be mandatory. The committee does not discuss those issues here.

invasive cervical cancer, vaccine efficacy, age at vaccination and vaccine coverage, the specificity and sensitivity of cervical cancer screening tests, vaccination costs (including the costs of the patients' time), the costs of cervical cancer screening and treatment (including the costs of the patients' time), and costs associated with health-related quality of life. Sanders and Taira (2003) conducted sensitivity analyses for various factors, including the optimal vaccination age, universal vaccination of adolescent girls versus targeting high-risk girls, the probabilities of occurrence and progression of HPV, squamous cell lesions and cervical cancer, the probability of death, and the costs and quality of life with various health states. Elbasha et al. (2007) evaluated the cost effectiveness of different HPV vaccination strategies, including (1) routine HPV vaccination of females by 12 years of age, (2) routine vaccination of females and males by 12 years of age, (3) routine vaccination of females by 12 years of age in combination with catch-up vaccinations for females between 12 and 24 years of age who were not previously vaccinated, and (4) routine vaccination of females and males by 12 years of age in combination with catch-up vaccinations for females and males between 12 and 24 years of age who were not previously vaccinated. The authors used a dynamic model that included demographic and epidemiologic components. Groups were formed on the basis of age and the extent of sexual activity (low, medium, or high sexual activity). The epidemiologic component "simulates HPV transmission and the occurrence of CIN, cervical cancer, and external genital warts" (Elbasha et al., 2007, p. 29). Vaccine characteristics included the degree of protection and the duration of protection. Taira et al. (2004) considered ranges of HPV prevalence, infection rates, age groups, sexual activity, and sex in order to estimate the cost effectiveness of different vaccination programs, including programs that would and would not include males.

In 2007, after considering those analyses and other factors, the ACIP recommended a three-dose vaccination series for females 11 to 12 years of age and, as a catch-up, vaccination for females 13 to 26 years of age who had not previously been vaccinated (Markowitz et al., 2007).

Since that decision, the ACIP has made three additional or updated recommendations related to HPV vaccination. In 2010 the committee developed a recommendation for Cervarix, a second HPV vaccine (CDC, 2010a). At the same time, the ACIP issued guidance stating that males aged 9 to 36 may be given Gardasil (CDC, 2010b). In 2011 the ACIP replaced that 2009 recommendation for males, recommending instead the use of Gardasil in males aged 11 or 12 years and a catch-up vaccination in males aged 13 through 21 who had not been vaccinated (or who had not completed the three-dose series) (CDC, 2011b). It also stated that males aged 22 to 26 years may be vaccinated.

In October 2010 the ACIP adopted a new framework for developing its recommendations, the Grading of Recommendations, Assessment, Development and Evaluation (GRADE) approach (Ahmed et al., 2011). The key factors in that approach include "the balance of benefits and harms, type of evidence, values and preferences of the people affected, and health economic analyses" (Ahmed et al., 2011, p. 9171). In the GRADE approach, ACIP includes tables that summarize the evidence on which it basis its decision. The tables include a description of the strength and limitations of the body of evidence. Those tables are intended to "enhance the ACIP's decision-making process by making it more transparent, consistent and systematic" (Ahmed et al., 2011, p. 9171).

Decisions in the Face of Uncertainty: Lessons Learned

Vaccinations illustrate a number of challenges experienced by those making decisions in the face of uncertainty. Viruses are "moving targets" that evolve and change, often requiring major policy decisions to be made in the absence of a full knowledge of the characteristics of future viruses. Faced with a possible pandemic in 1976, the government embarked on a vaccination campaign that has often been criticized. The lessons learned from that decision include the need for deliberative problem formulation, the importance of having different options, and the value of considering the potential effectiveness of those options in advance (for example, shutting down schools to control the spread of influenza). Other lessons from this case study are the importance of implementing an iterative approach in the presence of uncertainty, considering the feasibility of a plan, the importance of preparing for media communications, the importance of credibility, and the value of questioning knowledge and considering uncertainties in knowledge (Neustadt and Fineberg, 1978).

CDC's ACIP was faced with making recommendations about whom and when to vaccinate with a newly available, but expensive, vaccine against HPV that could prevent cervical cancer. A number of very detailed cost-effectiveness analyses were available that provided the ACIP with estimates of the benefits and costs associated with a number of different vaccination programs, assuming different levels of benefit. Those extensive analyses—which included analyses of a number of uncertainties—allowed ACIP to consider the effects of different scenarios and the range of benefits and costs under those scenarios using different estimates of vaccine effectiveness. ACIP's decision would potentially affect the medical care of the entire U.S. adolescent population. Having the detailed analyses showing ranges of cost effectiveness provided the ACIP with the evidence it needed to make a decision.

U.S. PREVENTIVE SERVICES TASK FORCE RECOMMENDATIONS

At the level of individual patients, physicians often make decisions or recommendations about preventive or therapeutic interventions with very uncertain information. They do so based on experiences from their training, advice from their professional societies, and information from government agencies and relevant private industries and by keeping current with the published medical literature and knowing their patients. In contrast to the population-based decisions that EPA makes, the decisions that physicians make or the advice they give is for one patient at a time, and the evidentiary basis for that advice typically comes from studies in populations and expert systematic reviews of the published literature, which may have different applicability to different patients. There is often conflicting evidence about the safety and effectiveness of the interventions. For example, although the U.S. Preventive Services Task Force (USPSTF) does not conduct quantitative uncertainty analysis when making its recommendations, the language of its recommendations qualitatively describes the level of uncertainty in the evidence (USPSTF, 2008). The USPSTF conclusions have significant policy implications. The role of certainty is explicitly stated in the conclusions and in the communication of the scientific evidence, providing an example of how a qualitative assessment and the description of uncertainties in an evidence base can provide important information to a decision maker. For example, the most strongly worded conclusion—for preventive interventions receiving a grade of A—reads, "The USPSTF recommends the intervention. There is high certainty that the net benefit is substantial." The Grade B conclusion reads, "The USPSTF recommends the service. There is high certainty that the net benefit is moderate or there is moderate certainty that the net benefit is moderate to substantial" (USPSTF, 2008). Unfortunately, for many interventions there is no clinical guidance given because the evidence is confusing or inadequate.

Decisions in the Face of Uncertainty: Lessons Learned

The USPSTF guidelines demonstrate how, for some types of decisions, simple, qualitative descriptions of the uncertainty can be helpful for the decisions.

KEY FINDINGS

- Some agencies, such as FSIS and FDA, have conducted quantitative uncertainty analyses on public health estimates. They use tools and techniques, such as probabilistic analysis and Monte Carlo simulations, that are similar to those that EPA has used.

- A phased or iterative approach to regulations, as occurred with the implementation of smoking bans, can allow for the collection of data, including data for economic analyses, that can decrease uncertainty.
- Regulations that are likely to be met with opposition, such as smoking bans and mandatory vaccinations, are well served to engage stakeholders in problem formulation and uncertainty analysis early to ensure that uncertainties are well understood and to ensure that stakeholders' information needs are met.
- Uncertainty analyses, such as those conducted in the assessment of *L. monocytogenes*, can characterize heterogeneity and its consequences and provide decision makers with the information to decide upon regulations and risk-mitigation options that target the public health goal in the most effective manner.
- Well-planned, decision-driven modeling of uncertainty, such as was conducted in the BSE risk assessment, can provide information about the likelihood of different regulatory options decreasing the risks to the public.
- Decisions must sometimes be made quickly in the face of large or deep uncertainty, such as was the case with the threatened pandemic influenza in 1976 and with melamine in the food supply in 2007 and 2008. Under such circumstances, probabilistic models are not typically available to help with decision making. The analysis of scenarios can be useful under such circumstances, and iterative management approaches can avoid mistakes that are costly either in terms of resources or to the reputation of and trust in the agency.
- Detailed economic analyses that outline the ranges of likely cost-effectiveness of different scenarios, such as those conducted for HPV vaccinations, can provide the evidence needed to make a decision.
- As demonstrated by FDA's activities around its decision on Avandia, making public the uncertainty from scientific disagreements—and even the disagreements among agency scientists—can increase the transparency and the public understanding of a decision. That decision also highlights how quantitative uncertainty analyses are not always needed to make an informed decision.
- All of these examples provide some characterization of uncertainty—some quantitative, some qualitative, some using safety factors—some in public health factors only and some in costs and economic impacts as well. These examples show the wide range of approaches that can be taken and provide some indication of the situations in which each approach may be appropriate. These examples also

show how uncertainty analysis is typically focused on public health impact estimates and demonstrate the few examples of assessing uncertainty in other factors. The tools and techniques used are typically those that EPA is already using.

REFERENCES

Ahmed, F., J. L. Temte, D. Campos-Outcalt, and H. J. Schünemann. 2011. Methods for developing evidence-based recommendations by the Advisory Committee on Immunization Practices (ACIP) of the U.S. Centers for Disease Control and Prevention (CDC). *Vaccine* 29(49):9171–9176.

Anderson, J. E., and S. Sansom. 2006. HIV testing among US women during prenatal care: Findings from the 2002 National Survey of Family Growth. *Maternal and Child Health Journal* 10(5):413–417.

ANRF (American Nonsmokers Rights Foundation). 2012. Overview list—How many smoke-free laws? http://www.no-smoke.org/pdf/mediaordlist.pdf (accessed July 10, 2012).

Baer, A., N. B. Kiviat, S. Kulasingam, C. Mao, J. Kuypers, and L. A. Koutsky. 2002. Liquid-based papanicolaou smears without a transformation zone component: Should clinicians worry? *Obstetrics and Gynecology* 99(6):1053–1059.

Barone-Adesi, F., L. Vizzini, F. Merletti, and L. Richiardi. 2006. Short-term effects of Italian smoking regulation on rates of hospital admission for acute myocardial infarction. *European Heart Journal* 27(20):2468–2472.

Bartecchi, C., R. N. Alsever, C. Nevin-Woods, W. M. Thomas, R. O. Estacio, B. B. Bartelson, and M. J. Krantz. 2006. Reduction in the incidence of acute myocardial infarction associated with a citywide smoking ordinance. *Circulation* 114(14):1490–1496.

Bernier, R. H., and E. K. Marcuse. 2005. Citizen voices on pandemic flu choices: A report of the Public Engagement Pilot Project on Pandemic Influenza (PEPPI). http://ppc.unl.edu/publications/documents/PEPPPI_FINALREPORT_DEC_2005.pdf (accessed January 3, 2013).

Bero, L. A., T. Montini, K. Bryan-Jones, and C. Mangurian. 2001. Science in regulatory policy making: Case studies in the development of workplace smoking restrictions. *Tobacco Control* 10(4):329–336.

Bonanni, P. 1999. Demographic impact of vaccination: A review. *Vaccine* 17(Suppl 3):S120–S125.

Borland, R., H. H. Yong, M. Siahpush, A. Hyland, S. Campbell, G. Hastings, K. M. Cummings, and G. T. Fong. 2006. Support for and reported compliance with smoke-free restaurants and bars by smokers in four countries: Findings from the International Tobacco Control (ITC) Four Country Survey. *Tobacco Control* 15(Suppl 3):34–41.

Bryan-Jones, K., and L. A. Bero. 2003. Tobacco industry efforts to defeat the Occupational Safety and Health Administration indoor air quality rule. *American Journal of Public Health* 93(4):585.

Cal EPA (California Environmental Protection Agency). 2005. Proposed identification of environmental tobacco smoke as a toxic air contaminant, part b: Health effects. California Environmental Protection Agency.

Carter, J. J., L. A. Koutsky, J. P. Hughes, S. K. Lee, J. Kuypers, N. Kiviat, and D. A. Galloway. 2000. Comparison of human papillomavirus types 16, 18, and 6 capsid antibody responses following incident infection. *Journal of Infectious Diseases* 181(6):1911–1919.

CDC (U.S. Centers for Disease Control and Prevention). 2010a. FDA licensure of bivalent human papillomavirus vaccine (HPV2, cervarix) for use in females and updated HPV vaccination recommendations from the Advisory Committee on Immunization Practices (ACIP). *Morbidity and Mortality Weekly Report* 59(20):626–629.

———. 2010b. FDA licensure of quadrivalent human papillomavirus vaccine (HPV4, gardasil) for use in males and guidance from the Advisory Committee on Immunization Practices (ACIP). *Morbidity and Mortality Weekly Report* 59(20):630–632.

———. 2011a. Multistate outbreak of listeriosis linked to whole cantaloupes from Jensen Farms, Colorado. http://www.cdc.gov/listeria/outbreaks/cantaloupes-jensen-farms/120811/index.html (accessed May 21, 2012).

———. 2011b. Recommendations on the use of quadrivalent human papillomavirus vaccine in males—Advisory Committee on Immunization Practices (ACIP), 2011. *Morbidity and Mortality Weekly Report* 60(50):1697–1728.

Cesaroni, G., F. Forastiere, N. Agabiti, P. Valente, P. Zuccaro, and C. A. Perucci. 2008. Effect of the Italian smoking ban on population rates of acute coronary events. *Circulation* 117(9):1183–1188.

Cohen, J. T. 2006. Harvard risk assessment of bovine spongiform encephalopathy update. Phase IA: Supplemental simulation results. Harvard Center for Risk Analysis. http://www.fsis.usda.gov/PDF/BSE_Risk_Assess_Report_2006.pdf (accessed January 2, 2013).

Cohen, J. T., and G. M. Gray. 2005. Harvard risk assessment of bovine spongiform encephalopathy update. Harvard Center for Risk Analysis. http://www.fsis.usda.gov/PDF/BSE_Risk_Assess_Report_2005.pdf (accessed January 2, 2013).

Cohen, J. T., K. Duggar, G. M. Gray, and S. Kreindel. 2003a. Evaluation of the potential for bovine spongiform encephalopathy in the United States: Report to the U.S. Department of Agriculture (revised October 2003). *Harvard Center for Risk Analysis*. http://www.agcenter.com/mad%20cow/Harvard%20Study%20on%20madcow[1].pdf (accessed January 2, 2013).

Cohen, J. T., K. Duggar, G. M. Gray, S. Kreindel, H. Gubara, T. HabteMariam, D. Oryang, and B. Tameru. 2003b. A simulation model for evaluating the potential for spread of bovine spongiform encephalopathy to animals or to people. In *Prions and mad cow disease*, edited by B. K. Nunnally and I. S. Krull. New York: Marcel Dekker. Pp. 53–111.

Darnton-Hill, I., and R. Nalubola. 2002. Fortification strategies to meet micronutrient needs: Successes and failures. *Proceedings of the Nutrition Society* 61(02):231–241.

Dunne, E. F., M. Sternberg, L. E. Markowitz, G. McQuillan, D. Swan, S. Patel, and E. R. Unger. 2011. Human papillomavirus (HPV) 6, 11, 16, and 18 prevalence among females in the United States—National Health and Nutrition Examination Survey, 2003–2006: Opportunity to measure HPV vaccine impact? *Journal of Infectious Disease* 204(4):562–565.

Elbasha, E. H., E. J. Dasbach, and R. P. Insinga. 2007. Model for assessing human papillomavirus vaccination strategies. *Emerging Infectious Diseases* 13(1):28. http://wwwnc.cdc.gov/eid/article/13/1/06-0438.htm (accessed January 2, 2013).

EPA (U.S. Environmental Protection Agency). 1992. Respiratory health effects of passive smoking: Lung cancer and other disorders. EPA/600/6-90/006F. Washington, DC: Environmental Protection Agency.

FDA (U.S. Food and Drug Administration). 1998. *Guidance for industry: Providing clinical evidence of effectiveness for human drugs and biological products*. Washington, DC: Food and Drug Administration. http://www.fda.gov/downloads/Drugs/GuidanceComplianceRegulatoryInformation/Guidances/ucm078749.pdf (accessed January 3, 2013).

———. 2003. *Reducing the risk of Listeria monocytogenes. FDA/CDC 2003 update of the Listeria Action Plan*. http://www.foodsafety.gov/~dms/lmr2plan.html#update (accessed March 3, 2012).

———. 2005. *Guidance for industry: Premarketing risk assessment.* http://www.fda.gov/downloads/RegulatoryInformation/Guidances/ucm126958.pdf%20 (accessed January 3, 2013).

———. 2007. Summary minutes of the Joint Meeting of the Endocrinologic and Metabolic Drugs Advisory Committee and the Drug Safety and Risk Management Advisory Committee held on July 30, 2007. http://www.fda.gov/ohrms/dockets/ac/07/minutes/2007-4308m1-final.pdf (accessed January 2, 2013).

———. 2008a. *Compliance policy guide—guidance for FDA staff, Sec. 555.320: Listeria monocytogenes draft guidance.* Washington, DC: U.S. Food and Drug Administration. http://www.fda.gov/OHRMS/DOCKETS/98fr/FDA-2008-D-0058-GDL.pdf (acessed January 3, 2013).

———. 2008b. "Dear colleague" letter to the United States food manufacturing industry, regarding melamine. http://www.fda.gov/Food/FoodSafety/FoodContaminantsAdulteration/ChemicalContaminants/Melamine/ucm164514.htm (accessed January 3, 2013).

———. 2008c. Guidance for industry: Guide to minimize microbial food safety hazards of fresh-cut fruits and vegetables. http://www.fda.gov/food/guidancecomplianceregulatoryinformation/guidancedocuments/produceandplanproducts/ucm064458.htm (accessed January 2, 2013).

———. 2008d. Interim safety and risk assessment of melamine and its analogues in food for humans [a]. http://www.fda.gov/Food/FoodSafety/FoodContaminantsAdulteration/ChemicalContaminants/Melamine/ucm164522.htm (accessed January 3, 2013).

———. 2008e. Update: Interim safety and risk assessment of melamine and its analogues in food for humans. http://www.fda.gov/Food/FoodSafety/FoodContaminantsAdulteration/ChemicalContaminants/Melamine/ucm164520.htm (accessed January 3, 2013).

———. 2010a. FDA drug safety communication: Ongoing review of Avandia (rosiglitazone) and cardiovascular safety. http://www.fda.gov/Drugs/DrugSafety/PostmarketDrugSafetyInformationforPatientsandProviders/ucm201418.htm (accessed January 2, 2013).

———. 2010b. FDA significantly restricts access to the diabetes drug Avandia. http://www.fda.gov/Drugs/DrugSafety/PostmarketDrugSafetyInformationforPatientsandProviders/ucm226956.htm (accessed January 2, 2013).

———. 2010c. FDA significantly restricts access to the diabetes drug Avandia: Makes regulatory decisions on record and tide trials. http://www.fda.gov/NewsEvents/Newsroom/PressAnnouncements/ucm226975.htm (accessed January 3, 2013).

———. 2011a. Raw Milk Misconceptions and the Danger of Raw Milk Consumption. http://www.fda.gov/Food/FoodSafety/Product-SpecificInformation/MilkSafety/ConsumerInformationAboutMilkSafety/ucm247991.htm (accessed January 17, 2013).

———. 2011b. Advisory committees: Critical to the FDA's product review process. http://www.fda.gov/Drugs/ResourcesForYou/Consumers/ucm143538.htm (accessed January 2, 2013).

———. 2011c. Environmental assessment: Factors potentially contributing to the contamination of fresh whole cantaloupe implicated in a multi-state outbreak of listeriosis. http://www.fda.gov/Food/FoodSafety/FoodborneIllness/ucm276247.htm (accessed January 2, 2013).

———. 2012a. Alert: The basics. http://www.fda.gov/Food/FoodDefense/ToolsResources/ucm296009.htm (accessed January 3, 2013).

———. 2012b. Guidance drug safety information—FDA's communication to the public. http://www.fda.gov/downloads/Drugs/GuidanceComplianceRegulatoryInformation/Guidances/UCM295217.pdf (accessed January 3, 2013).

———. 2012c. Guidance for industry and Food and Drug Administration staff. Factors to consider when making benefit-risk determinations in medical device premarketing approval and de novo classifications. http://www.fda.gov/downloads/MedicalDevices/DeviceRegulationandGuidance/GuidanceDocuments/UCM296379.pdf (accessed January 3, 2013).

FDA and FSIS (Food Safety and Inspection Service). 2003a. Interpretive summary: Quantitative assessment of relative risk to public health from foodborne *Listeria monocytogenes* among selected categories of ready-to-eat foods. http://www.fda.gov/downloads/Food/ScienceResearch/ResearchAreas/RiskAssessmentSafetyAssessment/UCM197329.pdf (accessed January 3, 2013).

———. 2003b. Quantitative assessment of relative risk to public health from foodborne *Listeria monocytogenes* among selected categories of ready-to-eat foods. http://www.fda.gov/downloads/Food/ScienceResearch/ResearchAreas/RiskAssessmentSafetyAssessment/UCM197330.pdf (accessed January 3, 2013).

———. 2007. Interim Melamine and Analogues Safety/Risk Assessment. http://www.fda.gov/Food/FoodSafety/FoodContaminantsAdulteration/ChemicalContaminants/Melamine/ucm164658.htm (accessed January 3, 2013).

Fischhoff, B., N. T. Brewer, and J. S. Downs (Eds.). 2011. *Communicating risks and benefits: An evidence-based user's guide*. Washington, DC: Food and Drug Administration.

FSIS (Food Safety and Inspection Service). 2003. Control of *Listeria monocytogenes* in ready-to-eat meat and poultry products: Final rule. Federal Register, June 6, 2003 (Volume 68, Number 109. http://www.fsis.usda.gov/oppde/rdad/frpubs/97-013f.htm (accessed January 3, 2013).

———. 2007. Disposition of hogs and chickens from farms identified as having received pet food scraps contaminated with melamine and melamine-related compounds and offered for slaughter. FSIS, Docket No. 2007-0018.

FSIS and USDA (U.S. Department of Agriculture). 2004. Preliminary analysis of interim final rules and an interpretive rule to prevent the BSE agent from entering the U.S. food supply. http://www.fsis.usda.gov/oppde/rdad/frpubs/03-025N/BSE_Analysis.pdf (accessed January 2, 2013).

———. 2007. Economic analysis: Final regulatory impact analysis. Final rule. FSIS Docket No. 03-025F.

Glantz, S. A., and A. Charlesworth. 1999. Tourism and hotel revenues before and after passage of smoke-free restaurant ordinances. *Journal of the American Medical Association* 281(20):1911–1918.

Glantz, S. A., and L. R. Smith. 1994. The effect of ordinances requiring smoke-free restaurants on restaurant sales. *American Journal of Public Health* 84(7):1081–1085.

———. 1997. The effect of ordinances requiring smoke-free restaurants and bars on revenues: A follow-up. *American Journal of Public Health* 87(10):1687–1693.

Goldie, S. J., M. Kohli, D. Grima, M. C. Weinstein, T. C. Wright, F. X. Bosch, and E. Franco. 2004. Projected clinical benefits and cost-effectiveness of a human papillomavirus 16/18 vaccine. *Journal of the National Cancer Institute* 96(8):604–615.

GRADE Working Group. 2012a. Organizations that have endorsed or that are using GRADE. http://www.gradeworkinggroup.org/society/index.htm (accessed June 13, 2012).

———. 2012b. Welcome. http://www.gradeworkinggroup.org/index.htm (accessed June 23, 2012).

Hackshaw, A. K., M. R. Law, and N. J. Wald. 1997. The accumulated evidence on lung cancer and environmental tobacco smoke. *British Medical Journal* 315(7114):980–988.

Hariri, S., E. R. Unger, M. Sternberg, E. F. Dunne, D. Swan, S. Patel, and L. E. Markowitz. 2011. Prevalence of genital human papillomavirus among females in the United States, the National Health and Nutrition Examination Survey, 2003–2006. *Journal of Infectious Diseases* 204(4):566–573.
Harper, D. M., E. L. Franco, C. Wheeler, D. G. Ferris, D. Jenkins, A. Schuind, T. Zahaf, B. Innis, P. Naud, N. S. De Carvalho, C. M. Roteli-Martins, J. Teixeira, M. M. Blatter, A. P. Korn, W. Quint, and G. Dubin. 2004. Efficacy of a bivalent L1 virus-like particle vaccine in prevention of infection with human papillomavirus types 16 and 18 in young women: A randomised controlled trial. *The Lancet* 364(9447):1757–1765.
Harro, C. D., Y.-Y. S. Pang, R. B. S. Roden, A. Hildesheim, Z. Wang, M. J. Reynolds, T. C. Mast, R. Robinson, B. R. Murphy, R. A. Karron, J. Dillner, J. T. Schiller, and D. R. Lowy. 2001. Safety and immunogenicity trial in adult volunteers of a human papillomavirus 16 L1 virus-like particle vaccine. *Journal of the National Cancer Institute* 93(4):284–292.
HHS (U.S. Department of Health and Human Services). 1986. *The health consequences of involuntary smoking: A report of the Surgeon General.* Washington, DC: U.S. Department of Health and Human Services.
HHS, CDC, Coordinating Center for Health Promotion, National Center for Chronic Disease Prevention and Health Promotion, and Office on Smoking and Health. 2006. *The health consequences of involuntary exposure to tobacco smoke: A report of the surgeon general.* Washington, DC: U.S. Department of Health and Human Services.
Hyland, A., K. M. Cummings, and E. Nauenberg. 1999. Analysis of taxable sales receipts: Was New York City's Smoke-Free Air Act bad for restaurant business? *Journal of Public Health Management and Practice* 5(1):14–21.
IOM (Institute of Medicine). 2004. *Advancing prion science: Guidance for the National Prion Research Program.* Edited by R. Erdtmann and L. B. Sivitz. Washington, DC: The National Academies Press.
———. 2006. *Modeling community containment for pandemic influenza: A letter report.* Washington, DC: The National Academies Press.
———. 2010a. *Secondhand smoke exposure and cardiovascular effects: Making sense of the evidence.* Washington, DC: The National Academies Press.
———. 2010b. *Women's health research: Progress, pitfalls, and promise.* Washington, DC: The National Academies Press.
———. 2012a. *Ethical and scientific issues in studying the safety of approved drugs.* Washington, DC: The National Academies Press.
———. 2012b. *Ranking vaccines: A prioritization framework: Phase I: Demonstration of concept and a software blueprint.* Edited by G. Madhavan, K. Sangha, C. Phelps, D. Fryback, T. Lieu, R. M. Martinez, and L. King. Washington, DC: The National Academies Press.
Juster, H. R., B. R. Loomis, T. M. Hinman, M. C. Farrelly, A. Hyland, U. E. Bauer, and G. S. Birkhead. 2007. Declines in hospital admissions for acute myocardial infarction in New York State after implementation of a comprehensive smoking ban. *American Journal of Public Health* 97(11):2035–2039.
Kelly, B. C., J. D. Weiser, and J. T. Parsons. 2009. Smoking and attitudes on smoke-free air laws among club-going young adults. *Social Work and Public Health* 24(5):446–453.
Khuder, S. A., S. Milz, T. Jordan, J. Price, K. Silvestri, and P. Butler. 2007. The impact of a smoking ban on hospital admissions for coronary heart disease. *Preventive Medicine* 45(1):3–8.
Law, M. R., J. Morris, and N. J. Wald. 1997. Environmental tobacco smoke exposure and ischaemic heart disease: An evaluation of the evidence. *Bristish Medical Journal* 315(7114):973–980.
Lemstra, M., C. Neudorf, and J. Opondo. 2008. Implications of a public smoking ban. *Canadian Journal of Public Health* 99(1):62–65.

Mangurian, C. V., and L. A. Bero. 2000. Lessons learned from the tobacco industry's efforts to prevent the passage of a workplace smoking regulation. *American Journal of Public Health* 90(12):1926-1930.

Markowitz, L. E., E. Dunne, M. Saraiya, H. Lawson, H. Chesson, and E. Unger. 2007. Quadrivalent human papillomavirus vaccine. *Morbidity and Mortality Weekly Report* 56(RR02):1-24.

Miller, C., M. Wakefield, S. Kriven, and A. Hyland. 2002. Evaluation of smoke-free dining in South Australia: Support and compliance among the community and restauranteurs. *Australia and New Zealand Journal of Public Health* 26(1):38-44.

Muggli, M. E., R. D. Hurt, and D. D. Blanke. 2003. Science for hire: A tobacco industry strategy to influence public opinion on secondhand smoke. *Nicotine and Tobacco Research* 5(3):303-314.

Neustadt, R. E., and H. V. Fineberg. 1978. *The swine flu affair: Decision-making on a slippery disease.* Washington, DC: National Academy Press.

Nissen, S. E., and K. Wolski. 2007. Effect of rosiglitazone on the risk of myocardial infarction and death from cardiovascular causes. *New England Journal of Medicine* 356(24):2457-2471.

NRC (Nuclear Regulatory Commission). 2012. Risk assessment in regulation. http://www.nrc.gov/about-nrc/regulatory/risk-informed.html (accessed January 3, 2013).

NTP (National Toxicology Program). 2011. *Report on carcinogens, twelfth edition.* Washington, DC: U.S. Department of Health and Human Services, Public Health Service, National Toxicology Program. http://ntp.niehs.nih.gov/ntp/roc/twelfth/roc12.pdf (accessed January 2, 2013).

Ong, E. K., and S. A. Glantz. 2000. Tobacco industry efforts subverting International Agency for Research on Cancer's second-hand smoke study. *Lancet* 355(9211):1253-1259.

Oreskes, N., and E. M. Conway. 2010. *Merchants of doubt: How a handful of scientists obscured the truth on issues from tobacco smoke to global warming.* New York: Bloomberg Press.

Pell, J. P., S. Haw, S. Cobbe, D. E. Newby, A. C. H. Pell, C. Fischbacher, A. McConnachie, S. Pringle, D. Murdoch, and F. Dunn. 2008. Smoke-free legislation and hospitalizations for acute coronary syndrome. *New England Journal of Medicine* 359(5):482-491.

Petter, A., K. Heim, M. Guger, A. Ciresa-König, N. Christensen, M. Sarcletti, U. Wieland, H. Pfister, R. Zangerle, and R. Höpfl. 2000. Specific serum IGG, IGM and IGA antibodies to human papillomavirus types 6, 11, 16, 18 and 31 virus-like particles in human immunodeficiency virus-seropositive women. *Journal of General Virology* 81(3):701-708.

Pursell, L., S. Allwright, D. O'Donovan, G. Paul, A. Kelly, B. J. Mullally, and M. D'Eath. 2007. Before and after study of bar workers' perceptions of the impact of smoke-free workplace legislation in the republic of Ireland. *BMC Public Health* 7:131.

Sanders, G. D., and A. V. Taira. 2003. Cost effectiveness of a potential vaccine for human papillomavirus. *Emerging Infectious Diseases* 9(1):37-48.

Sargent, R. P., R. M. Shepard, and S. A. Glantz. 2004. Reduced incidence of admissions for myocardial infarction associated with public smoking ban: Before and after study. *British Medical Journal* 328(7446):977-980.

Scallan, E., R. M. Hoekstra, F. J. Angulo, R. V. Tauxe, M. A. Widdowson, S. L. Roy, J. L. Jones, and P. M. Griffin. 2011. Foodborne illness acquired in the United States—Major pathogens. *Emerging and Infectious Diseases* 17(1):7-15.

Sciacca, J. P., and M. I. Ratliff. 1998. Prohibiting smoking in restaurants: Effects on restaurant sales. *American Journal of Health Promotion* 12(3):176-184.

Scollo, M., A. Lal, A. Hyland, and S. Glantz. 2003. Review of the quality of studies on the economic effects of smoke-free policies on the hospitality industry. *Tobacco Control* 12(1):13-20.

Seo, D. C., and M. R. Torabi. 2007. Reduced admissions for acute myocardial infarction associated with a public smoking ban: Matched controlled study. *Journal of Drug Education* 37(3):217–226.

Swaminathan, B., and P. Gerner–Smidt. 2007. The epidemiology of human listeriosis. *Microbes and Infection* 9(10):1236–1243.

Taira, A. V., C. P. Neukermans, and G. D. Sanders. 2004. Evaluating human papillomavirus vaccination programs. *Emerging Infectious Diseases* 10(11):1915–1923.

Tang, H., D. W. Cowling, J. C. Lloyd, T. Rogers, K. L. Koumjian, C. M. Stevens, and D. G. Bal. 2003. Changes of attitudes and patronage behaviors in response to a smoke-free bar law. *American Journal of Public Health* 93(4):611–617.

Tong, E. K., and S. A. Glantz. 2007. Tobacco industry efforts undermining evidence linking secondhand smoke with cardiovascular disease. *Circulation* 116(16):1845–1854.

U.S. Surgeon General's Advisory Committee on Smoking and Health. 1964. *Smoking and health. Report of the Advisory Committee to the Surgeon General of the Public Health Service.* Public Health Service Publication No. 1103 United States. Public Health Service. Office of the Surgeon General.

USDA (U.S. Department of Agriculture)/FSIS (Food Safety and Inspection Service) and EPA (U.S. Environmental Protection Agency). 2012. Microbial risk assessment guideline: Pathogenic organisms with focus on food and water. FSIS Publication No. USDA/FSIS/2012-001; EPA Publication No. EPA/100/J12/001.

USPSTF (U.S. Preventive Services Task Force). 2008. Grade definitions. http://uspreventiveservicestaskforce.org/uspstf/grades.htm#post (accessed January 2, 2013). http://www.fsis.usda.gov/PDF/Microbial_Risk_Assessment_Guideline_2012-001.pdf (accessed January 3, 2013).

Vasselli, S., P. Papini, D. Gaelone, L. Spizzichino, E. De Campora, R. Gnavi, C. Saitto, N. Binkin, and G. Laurendi. 2008. Reduction incidence of myocardial infarction associated with a national legislative ban on smoking. *Minerva Cardioangiologica* 56(2):197–203.

WHO (World Health Organization). 2006. *Building blocks for tobacco control: A handbook.* Geneva: World Health Organization.

———. 2013. Microbiological risks publications. http://www.who.int/foodsafety/publications/micro/en (accessed January 3, 2013).

Woodcock, J. 2010. Memorandum from Janet Woodcock to NDA 021071 (dated September 22, 2010). Re: Decision on continued marketing of rosiglitazone (Avandia, Avandamet, Avandaryl). http://www.fda.gov/downloads/Drugs/DrugSafety/Postmarket DrugSafetyInformationforPatientsandProviders/UCM226959.pdf (accessed January 3, 2013).

5

Incorporating Uncertainty into Decision Making

As outlined in Chapter 1, the committee focused on the uncertainty in three types of factors that can play a role in the decisions of the U.S. Environmental Protection Agency (EPA): health, technological, and economic. Historically, uncertainties in health estimates have received the most attention (see Chapter 2). Uncertainties in technological and economic factors have received less attention (see Chapter 3). In this chapter, the committee presents a framework to help EPA incorporate uncertainty in the three factors into its decisions. Where possible, the committee incorporates the lessons from other public health agencies discussed in Chapter 4.

Science and Decisions: Advancing Risk Assessment (hereafter *Science and Decisions*) (NRC, 2009) recommended a three-phase decision-making framework consisting of problem-formulation, assessment, and management phases. Both *Science and Decisions* and the framework it suggests emphasize the need to do a better job of linking the assessment of health risks to the particular problem that EPA is facing and also emphasize the importance of stakeholder involvement in each stage.

In this chapter this committee begins by building on that three-phase framework, incorporating into that framework uncertainty in the three factors (health risk estimates, technology availability, and economics) that play a role in EPA's decisions. As with the framework from *Science and Decisions* (NRC, 2009) and other decision-making frameworks (see, for example, Gregory, 2011; Gregory et al., 1996; Spetzler, 2007), this committee's framework emphasizes the importance of interactions between decision makers, stakeholders, and analysts. The modified framework is presented in Figures 5-1a and 5-1b and is discussed below. After introducing the

148

PHASE I: PROBLEM FORMULATION AND SCOPING
- What problems are associated with existing environmental conditions?
- If existing conditions appear to pose a threat to human health, what options exist for altering those conditions?
- What is the legal context of the decisions and, therefore, what factors should EPA consider in its decisions?
- Under the given decision context, what scientific or technical assessments are necessary to evaluate possible management decisions
- What are the potential uncertainties associated with the relevant factors for decision making?
- What populations should be considered in assessments?
- Should a quantitative or qualitative assessment of those uncertainties be conducted
- Is there research that could be conducted within an acceptable timeframe that would decrease any of the uncertainties?
- Are there any deep uncertainties related to the issue?

PHASE II: PLANNING AND CONDUCT OF ASSESSMENTS

Stage 1: Planning of the Assessments
- For the given decision context, what are the attributes of studies necessary to characterize the human health risks, technology, and economics that should be assessed for the proposed regulatory options?

Stage 2a: Human Health Risk Assessment

Stage 2b: Technology Availability Assessment

Stage 2c: Economic Analysis

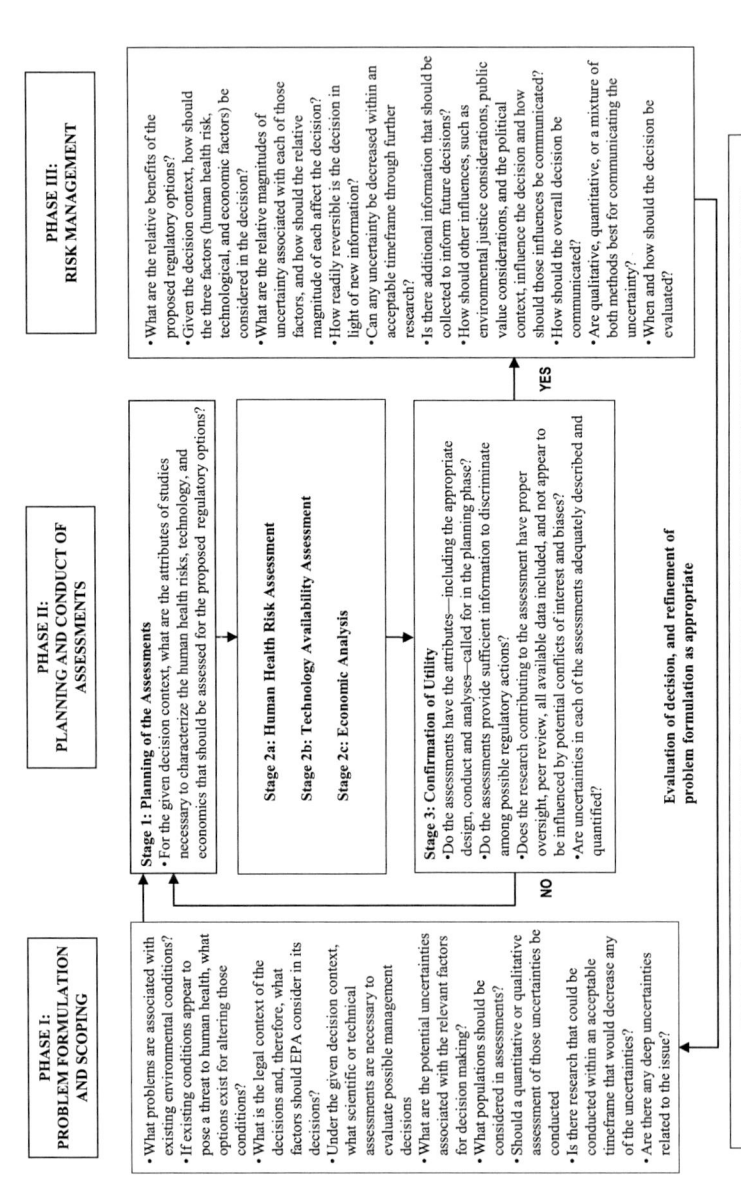

Stage 3: Confirmation of Utility
- Do the assessments have the attributes—including the appropriate design, conduct and analyses—called for in the planning phase?
- Do the assessments provide sufficient information to discriminate among possible regulatory actions?
- Does the research contributing to the assessment have proper oversight, peer review, all available data included, and not appear to be influenced by potential conflicts of interest and biases?
- Are uncertainties in each of the assessments adequately described and quantified?

Evaluation of decision, and refinement of problem formulation as appropriate

PHASE III: RISK MANAGEMENT
- What are the relative benefits of the proposed regulatory options?
- Given the decision context, how should the three factors (human health risk, technological, and economic factors) be considered in the decision?
- What are the relative magnitudes of uncertainty associated with each of those factors, and how should the relative magnitude of each affect the decision?
- How readily reversible is the decision in light of new information?
- Can any uncertainty be decreased within an acceptable timeframe through further research?
- Is there additional information that should be collected to inform future decisions?
- How should other influences, such as environmental justice considerations, public value considerations, and the political context, influence the decision and how should those influences be communicated?
- How should the overall decision be communicated?
- Are qualitative, quantitative, or a mixture of both methods best for communicating the uncertainty?
- When and how should the decision be evaluated?

FORMAL PROVISIONS FOR INTERNAL AND EXTERNAL STAKEHOLDER INVOLVEMENT DURING ALL PHASES

The involvement of decision makers, technical specialists, and other stakeholders in all phases of the process leading to decisions should in no way compromise the technical assessment of risk, which is conducted under its own standards and guidelines.

FIGURE 5-1a Framework for decision making.
SOURCE: Modified from NRC, 2009.

> **Stage 2a: Human Health Risk Assessment**
> *Hazard Assessment*
> • What are the adverse health effects associated with the agents of concern?
> • What are the uncertainties (qualitative or quantitative) associated with those estimates?
> *Exposure Assessment*
> • What exposures/doses are incurred by each population of interest under existing conditions?
> • How does each regulatory option affect existing conditions and resulting exposure/doses?
> • What are the uncertainties (qualitative or quantitative) associated with those estimates?
> *Risk Characterization*
> • What is the nature and magnitude of risk associated with existing conditions?
> • What risk decreases (benefits) are associated with each of the regulatory options?
> • Are any risks likely to be increased?
> • What significant uncertainties remain and what are their potential impact on management decisions?

> **Stage 2b: Assessment of Technology Availability**
> • What technologies are available?
> • What technologies are likely to soon become available?
> • Will a change in the regulation spur the development of new technologies?
> • What costs are associated with different technologies?
> • What are the expected decreases in emissions or exposures anticipated from the different technologies?
> • What uncertainties are associated with each of those estimates?

> **Stage 2c: Economic Analysis**
> • What are the costs associated with different regulatory options?
> • What are the benefits associated with different regulatory options?
> • What uncertainties are associated with each of those estimates?

FIGURE 5-1b[1] Considerations for each assessment during phase 2.

framework, the report then discusses different approaches to handling uncertainty in health, technological, and economic factors and describes ways that stakeholder engagement may be encouraged.

INCORPORATING UNCERTAINTY INTO A DECISION-MAKING FRAMEWORK

Problem Formulation

The need for EPA to make a regulatory decision might arise from concerns about a potential environmental hazard, a legal requirement to review an existing or potential environmental regulation, or concerns following a specific event, such as an oil spill or the siting of a new source of pollution. Regardless of why a decision is needed, when approaching a regulatory decision EPA should first identify and characterize the question

---

[1] Due to a production error, Figure 5-1b was inadvertently left out of the prepublication copy of this report.

or problem that underlies the regulatory decision. In other words, it first needs to perform a problem formulation and scoping.

Science and Decisions (NRC, 2009) highlights the importance of the *problem formulation* phase, which includes identifying the environmental concerns, planning and determining the scope of decision making, and identifying potential regulatory options and the criteria for selecting among those options. This committee agrees with the earlier report that planning for a risk assessment and anticipating issues in advance are key to conducting useful and high-quality assessments (such as assessments of human health risks, cost–benefit assessments, and assessments of technology availability), and the committee further emphasizes the importance of identifying uncertainties that affect the decision and determining how those uncertainties should be assessed and considered in the decision-making process. Identifying potential regulatory actions during this first phase will facilitate identifying the uncertainty surrounding the consequences of the regulatory actions, in order to plan any assessments of those uncertainties. Although not all stakeholders will necessarily agree with a regulatory decision—some will refuse to support any increase in regulation, for example, while others will refuse to support any decrease in regulation—an enhanced problem formulation will help to ensure that the different participants are aware of the different perspectives and that many of the potential uncertainties are identified. Identifying and characterizing the problem and potential regulatory options as well as planning for the uncertainty analysis are discussed below.

Identifying and Characterizing the Problem and Potential Regulatory Options

As discussed in *Understanding Risk* (NRC, 1996), the assessments of health risks and other factors should be decision driven, that is, driven by the context of the decision. Stakeholders, however, often have different views and perspectives on what problem underlies or caused the need for a decision, what information is available and should be considered when making a decision, and what uncertainties could affect a decision (Koppenjan and Klijn, 2004). Those different views and perspectives could, in part, determine the most appropriate way to assess the factors in the decision (such as health risks, costs, and technology). All participants, therefore, need to be aware of and understand the views and perspectives of others, as well as have a common understanding of the problem to be addressed, the purpose of the assessments, and the potential regulatory options.

Complex decisions that affect multiple stakeholders benefit from a formal process that ensures that the problem and the solutions are adequately characterized and agreed upon by all parties. Problem-structuring methods for unstructured problems which—like many of the problems EPA faces, have multiple actors and perspectives, incommensurable or conflicting interests, important intangibles, and key uncertainties—provide a "way of representing the situation . . . that will enable participants to clarify their predicament, converge on a potentially actionable mutual problem or issue within it, and agree [on] commitments that will at least partially resolve it" (Mingers and Rosenhead, 2004, p. 531). The interaction among stakeholders that occurs with problem-structuring methods typically helps not only to build a consensus about a problem, but also to build social trust (see Chapter 6 for further discussion of social trust).

A small but growing literature from operations research provides guidance on problem structuring (see Gregory, 2011; Gregory and Keeney, 2002; Gregory et al., 1996; Hammond et al., 1984; Rosenhead, 1996; von Winterfeldt and Fasolo, 2009). According to this literature, in order to structure a problem one should (1) focus on the decision, that is, on the policy or regulatory choices and objectives; (2) maintain a broad perspective, that is, do not narrow down decision alternatives or objectives too early; and (3) involve a broad range of stakeholders to assist in identifying alternatives and objectives, thereby creating a legitimate framing of the policy or regulatory problem.

For environmental policy and regulation, the policy or regulatory objectives could include health risk reduction, an improvement of the environment, minimizing direct implementation costs, minimizing indirect and long-term socioeconomic impacts, or identifying a solution that maximizes the net benefit. Some studies favor structuring the problem in terms of the net benefits (that is, the total benefits minus the total costs from, for example, health improvements) rather than in terms of the risk reduction (Stinnett and Mullahy, 1998).

The planning of assessments should include not only assessments of health risks and benefits, but also assessments of the other factors that might be considered in a decision, in particular, technological and economic factors. Keeney and Raiffa (1976) and Keeney (1996) discussed how to generate a comprehensive set of objectives, including identifying which direct and indirect costs should be considered part of the objectives. Garber and Phelps's (1992) work on the near equivalence of benefit–cost analysis and cost-effectiveness analysis, along with the method of cost acceptability curves (Fenwick et al., 2001), lead to a larger framework for analyzing uncertainty in both benefit–cost analysis and cost-effectiveness contexts. The metrics that will be used to measure the objectives should also be defined as part of this phase.

Planning for the Uncertainty Analysis

As can be seen in Figure 5-1a, planning for the analyses of uncertainty should begin during the problem-formulation phase. A major challenge is determining whether and how uncertainties should be quantified and how they should be taken into account in a regulatory decision. When considering how to analyze uncertainties, the type and complexity of the uncertainty analyses that are appropriate will depend on, among other things, the context of the decision (for example, if it is made in an emergency situation, the level of controversy and scientific disagreement around the decision, and whether the decision would be easily reversible), the nature of the risks and benefits (for example, if the human health risks involve minor adverse events, complex quantitative uncertainty analyses might not be warranted, whereas if they involve a fatal, nonreversible disease, such analyses might be warranted), the factors considered in the decision (for example, economic, technological, or social factors), and the type (for example, variability, model uncertainty, or deep uncertainty) and magnitude of the uncertainty. In particular, environmental statutes distinguish between decision contexts that are solely based on health considerations and those that consider technological feasibility or availability, cost–benefit trade-offs, or some combination of the three different types of considerations. It is important, therefore, that EPA identify in the problem-formulation phase of its decision-making process those factors it needs to consider in the decision and the nature or type of the uncertainty in those factors. That identification could involve providing a list of items that contribute to uncertainty, such as limited data, alternative models, or disagreements among the experts. For some factors, the process may include providing ranges of estimates from the literature or some preliminary representations of uncertainty, such as event trees, influence diagrams, or belief nets (see Box 5-1).

There is no "one-size-fits-all" approach for an agency to make decisions in the face of uncertainty, nor is a particular approach to uncertainty analysis appropriate for all decisions, but, in general, certain types of approaches and analyses lend themselves to certain types and sources of uncertainty. In Table 5-1, as a guide for EPA, the committee presents a typology of decision situations which indicates when different approaches to handling uncertainty might be appropriate. Those approaches are discussed in more detail in Appendix A.

The legal context determines which factors—health, technology, and economics—can be considered in EPA's decisions and, therefore, should be assessed (shown in the columns in Table 5-1). Each of those three factors can exhibit any or all of the three types of uncertainty to different extents (shown in the rows in Table 5-1), and each combination of factor and type of uncertainty lends itself to a different type of uncertainty analysis.

> **BOX 5-1**
> **Definitions of Preliminary Graphical**
> **Representations of Uncertainty**
>
> **Belief nets** "represent the causal and noncausal structure of inferences going from data and inference elements to the main hypothesis or event about which an inference is made" (von Winterfeldt, 2007).
>
> **Event trees** start with an initiating event and trace that to the "fault or problem event" (von Winterfeldt, 2007).
>
> **Influence diagrams** are graphical or visual representations of a decision situation. Conventionally, uncertain variables, decision nodes, and value nodes are shown in ellipses, rectangles, and rounded rectangles, respectively (von Winterfeldt, 2007).

While the regulatory context specifies which factors EPA can consider in making decisions, many of EPA's decisions will involve multiple types of uncertainty. EPA's plans for assessing uncertainty, therefore, will involve multiple analyses and approaches. The committee does not present all possible analytic approaches; rather it presents a number of approaches as a starting point to indicate how EPA should plan its analyses during the first phase of its decision-making process.

Looking across the columns in Table 5-1 at the legal or regulatory context, if the context is narrow—such as cases in which only health effects are taken into account (first column, Table 5-1)—then the approaches to uncertainty would typically be limited to versions of using safety or default factors (see Chapter 2 for further discussion); health risk analysis, including extreme value analysis; and scenario analysis, depending on the type or nature of the uncertainty. If technological availability—such as the best available or best practicable technology—can be considered (second column, Table 5-1), then health effects analyses can be combined with an assessment of the availability or practicability of the technological option, estimated using direct assessments or technological choice/risk analyses, to reduce health effects. If cost–benefit factors are allowed (third column, Table 5-1), appropriate analytic approaches include cost-effectiveness analysis, cost–benefit analysis, and multiattribute utility analysis. Cost-effectiveness, cost–benefit, multiattribute utility analysis, and decision analysis do not differ if the uncertainties are in variability and heterogeneity or in models and parameters. In the case of deep uncertainty, there is a shift to scenario analysis and robust decision-making tools, but in the case of cost–benefit factors, this analysis would include deep uncertainty about all factors

TABLE 5-1 Influence of the Type and Source of Uncertainty on Incorporating Uncertainty into a Decision

		REGULATORY CONTEXT: FACTORS CONSIDERED IN THE DECISION[a]		
		Health Effects Only	Technology Availability	Cost–Benefit

TYPE OF UNCERTAINTY	Variability and Heterogeneity[b]	• Use of safety or default factors (using statistics) if little or no data on uncertainties are available, and
		• Analysis of statistical distributions, including extreme value analysis if data are available
			• Using statistics for
			  o direct assessments, and
			  o technological choice/risk analysis
				• Using statistics for
				  o cost-effectiveness analysis,
				  o cost–benefit analysis, and
				  o multiattribute utility analysis

	Model and Parameter Uncertainty[c]	• If little or no data are available, using expert judgments and the use of safety or default factors, and
		• If data are available, using expert elicitation and analysis of probability distributions, including extreme value analysis
			• Using formal expert elicitation to assess technology availability, and
			• Using expert judgments for technology choice/risk analysis
				• Using expert judgments for
				  o cost-effectiveness analysis,
				  o cost–benefit analysis, and
				  o decision analysis

	Deep Uncertainty[d]	Scenario analysis and robust decision-making methods		

NOTES: The most appropriate methods to evaluate, analyze, or account for uncertainty often depend on the types and sources of uncertainty that are present. The columns of the matrix show which methods are typically appropriate for different regulatory contexts, that is, what factors environmental laws and executive orders require the Environmental Protection Agency (EPA) to consider in a given decision. The rows of the matrix show the methods that are often appropriate for heterogeneity and variability, model and parameter uncertainty, and deep uncertainty.

[a] The regulatory (or legal) context determines, to a large extent, what factors EPA considers in its regulatory decisions.

[b] The goal of assessing uncertainty from variability and heterogeneity is to identify different populations (health), technology and facilities (technology), or regulatory options (cost–benefit tradeoffs) and to estimate (with uncertainty) the magnitude of the differences among them.

[c] The goal of assessing model and parameter uncertainty is to estimate (with uncertainty) the effect of model choice and parameter values on assessments of health risks, technological factors, and the cost–benefit tradeoffs of different regulatory options.

[d] The goal is to identify deep uncertainties in the assessments, their potential effects on a decision, whether to conduct research to decrease the uncertainties, and when decisions should be revisited in light of those uncertainties. Both variability and heterogeneity, and model and parameter uncertainty, can be deep uncertainty.

considered in the analysis. If model or parameter uncertainty is present, expert judgments or elicitations can be helpful in estimating human health risks, technology availability, and costs and benefits.

Looking down the columns in Table 5-1 shows that different types of uncertainty lend themselves to different approaches to assessing and considering uncertainty. *Statistical methods* are appropriate for situations involving large amounts of data that allow uncertainty assessments by fitting standard probability distributions to data, that is, when uncertainties are primarily related to statistical variability and population heterogeneity. *Expert judgment techniques* or *safety* or *default factors* are needed when models and their parameters are uncertain and when data are sparse, for example, when the slope or shape of the dose–response function is uncertain or when extrapolation from animal data to humans is necessary. When facing deep uncertainties, probabilistic methods are more limited in use, and scenario analysis, sometimes coupled with robust decision-making methods, can help (see further discussion later in this chapter). Robust decision-making methods are those that provide acceptable outcomes for a range of possible scenarios, including pessimistic ones.

The goals in the assessing the different types of uncertainty are also different. For variability or heterogeneity (first row, Table 5-1), the goal of the assessment approach is to identify the subpopulations that are differentially affected, estimate the magnitude of the differences in results in the different subpopulations and the within-subpopulation variability, and assess the uncertainty in those estimates. For model and parameter uncertainty (second row, Table 5-1), the goal of the assessment approach is to compare results based on specifications with different functional forms, and one should compare simulations using different assumptions about the parameters depicting relationships between key explanatory variables and the dependent variables. For deep uncertainty (third row, Table 5-1), the goal or purpose of the assessment approach is fundamentally different; scenarios of various adverse outcomes should be described, and an assessment should be made as to whether a proposed solution can eliminate the risks of those outcomes occurring.

When accounting for uncertainty in a regulatory decision, each analysis or approach is associated with a set of *decision rules* that identify the "best" regulatory decision if the decision maker were to follow the recommendations resulting from the analysis. For example, a decision rule for a cost–benefit analysis would be to select the regulatory option with the highest net social benefit.

Further details about the specific approaches to assessing and considering uncertainty in decisions are presented later in this chapter.

Assessment

Once the decision makers, analysts, and stakeholders have a clear understanding of what assessments (human health risk assessments, economic analysis, or assessments of technology availability) are needed to inform a given decision, how those assessments should be conducted, and the uncertainties that need to be analyzed, the assessment phase begins. Assessment refers to the collection of data, modeling, and the estimation of impacts in order to determine how the regulatory options (including the status quo) perform with respect to the objectives specified in the problem-formulation stage (NRC, 2009). This is the factual part of the decision-making process and provides the analytic basis for the management phase, which involves evaluation, decision making, value-of-information analysis, and implementation.

The objective of the assessment phase is to analyze the available data or evidence and provide decision makers with the analyses in a way to inform the decision, including providing information about the uncertainties in the data and in the overall assessment. It is crucial that analysts do not lose sight of that objective when conducting uncertainty analyses. For example, they should not use extensive resources to analyze an uncertainty in a parameter or factor that has little relevance to the overall decision. It is also crucial that decision makers understand the implications of choices that analysts might make in the assessment process. For example, decision makers need to be aware of whether any default assumptions or models are embedded in an assessment and how those defaults might affect the assessment.

A main objective of EPA's regulatory decisions is to reduce adverse human health and environmental outcomes. Human health risk assessment is a well-understood and mature activity at EPA and other regulatory agencies. As described in *Risk Assessment in the Federal Government: Managing the Process* (NRC, 1983), it includes hazard identification (determining which health and environmental impacts are pertinent to the decision under consideration, with more specificity than the broader objectives specified in the first phase), an exposure assessment (assessing the levels of exposure to environmental agents), a dose–response assessment (a quantitative analysis of the effect of a unit change in exposure to particular environmental agents on specific health and environmental outcomes), and risk characterization (the health and environmental outcomes expected at a specific level of exposure to an environmental hazard). Human health risk assessment is conducted for the base case (outcomes at a future date if no change in regulation is implemented) and for one or more regulatory options. Human health risk assessment is the tool that decision makers use to predict the degree of health improvement or protection expected from a decrease in one or more exposures. Such risk assessments do not, however, indicate which intervention to use—that is, which is the best way to decrease exposure.

For the assessment phase, most previous NRC reports and EPA risk assessments have focused only on the assessment of health risks and their associated uncertainties. This committee, however, believes that the assessment phase should also include examinations of a number of nonhealth factors and their associated uncertainties. In particular, assessments should include technological factors and economic factors. The next section briefly describes where uncertainties can arise in the assessments of factors other than human health risks. For more details of assessments and assessment techniques for human health risks and for the various other factors, readers should refer back to Chapters 2 and 3, respectively.

It is worth noting that uncertainties are expressed as probabilities or probability distributions. While there is some discussion in the literature about the use of qualitative (verbal) vs. quantitative (numerical) expressions of probabilities (von Winterfeldt, 2007), most studies of environmental uncertainties use quantitative probabilities because they lend themselves to a wide array of statistical and other analyses. There are also different schools of thought about what these probabilities represent, including the classical or logical view, the frequentistic view, and the subjective or Bayesian view. Without taking a side in the debate among these schools of thought, the committee takes it for granted that probabilities are always based on logic, data, and judgment and that they should be informed and revised as new information is obtained.

Typically, EPA's decisions take into consideration the costs incurred by private parties as well as by public agencies. Private parties bear the cost of mitigation to reduce health and environmental impacts, while the public sector bears the costs of monitoring and enforcement as well as other costs. Both public and private costs are likely to be uncertain for several reasons. On the private side, the costs of mitigation alternatives are often uncertain. Technological development, whose outcome is often uncertain, may be required in many cases, and the uncertainties related to that development add to the uncertainty in the eventual technology costs. On the public side, there are choices to be made concerning the level of regulatory enforcement. An air standard can be enforced with more or less effort devoted to detection or prosecution; each choice implies expending a different amount of public resources. Changes in the level of enforcement are likely to lead to a change in the levels of benefits of the policy. For example, an unenforced standard may be of no benefit except perhaps to signal that some decision maker is sympathetic to a particular cause.

At the same time, public expenditures on enforcement may vary from those projected at the time the policy was implemented because individuals and firms in the private sector respond differently to the policy than originally anticipated. Given the very imperfect information that regulatory agencies have about the investments that private-sector organizations will make in response to regulatory rules, private-sector costs are likely to be subject to considerable variability and thus to uncertainty.

Regulatory options also need to be assessed with respect to nonenvironmental or health risk objectives. Even if the relevant statute requires that the regulation be based solely on health considerations, an analysis of other factors and their uncertainties can be useful for deciding among regulatory options that may have different associated costs, including such adverse consequences as loss of employment, or that are less or more feasible given current technologies or technologies that are likely to be developed in the foreseeable future.

The result of this state of affairs is that the assessments of both health risks and other factors are fraught with uncertainties. To help determine which assessments should explicitly include an uncertainty analysis, it is useful to first conduct a rudimentary, deterministic assessment using base-case values. In such an assessment it is assumed that model parameters and causal relationships are exact. For example, one might assume in the base-case scenario that a 10 percent increase in a pollution level leads to a 1 percent increase in a specific type of mortality. The sensitivity of the estimates of health risks and other estimates to changes in the base-case parameters and assumptions should then be explored. If the increase in exposure has a 95 percent likelihood of occurring, then the 10 percent increase in exposure would be associated with a mortality change in the range 0.5 to 1.5 percent. In sensitivity analysis we can explore mortality changes at specific values within this range. Realistically, the distribution is often not nearly as tight as the example implies. Because the final estimates may come from a chain of models, each of which has uncertainties, the potential errors in each link propagate through the chain, increasing the uncertainty of the estimated final effects.

Such sensitivity analysis serves two essential purposes. First, it helps identify those cases in which an analysis of uncertainty should be undertaken. For example, if there is a high likelihood that the credible interval[2] is narrow, then the actual magnitude of the effect is unlikely to change a decision about whether to conduct an analysis of uncertainty. This analytic decision depends in part on the extent to which conclusions are robust to changes in baseline assumptions and parameters (that is, to the extent that the conclusion is unchanged by results of the sensitivity analysis). If assuming an effect of 1.5 percent instead of an effect of 0.5 percent does not

---

[2] In this report the committee used the term *credible interval* when making a statement about a hypothesis given the data; the term *confidence interval* is used only when a statement is being made about the data, given a hypothesis.

change the decision, and if there is reasonably certainty that the confidence interval is correct, there is no reason to conduct an additional analysis. If, however, there is credible reason to suspect that the confidence interval may be much larger than that, it may be appropriate to carry out additional analysis in order to resolve the substantive issue. Second, to the extent that results are shown not to be robust, sensitivity analysis is helpful in identifying those assumptions and parameters that most influence the projected outcomes and therefore warrant further study. Although the above example assumes that there is only one parameter, typically there would be several parameters, each with associated confidence intervals, as well as parameters for interaction terms, all of which greatly complicate the problem.

Management

The previous sections have described potential analytic approaches for dealing with various types of uncertainty. However, applying these approaches in the context of risk management and environmental decision making is as much an art as a science. Risk management consists of evaluating the assessment results, making a decision, and implementing the decision, including monitoring the effects or outcomes from that decision or regulatory action. As is the case the problem-formulation and assessment phases, uncertainty plays an important role in this phase.

Stakeholder engagement is also an important aspect of risk management. As discussed elsewhere in this report, considering the input of stakeholders, including both the variability in their views and how the uncertainty in different factors could affect different stakeholders, is important in environmental regulatory decision making. Another important aspect of stakeholder engagement during the management phase is communication of the decision and the estimated health risks, cost, and other consequences of the decision along with the uncertainty in those estimates. Communication is discussed further in Chapter 6.

Evaluation of Assessment Results

Decision making is more than the formal evaluation of alternatives; decision makers must consider the results of the assessments and the uncertainty in those results in the context of additional informal and nonquantifiable aspects of the decision problem. Their decision should depend not only on the results of the assessments (for example, the human health, technological, or economic assessment), but also on interpretation of those results in the context of the decision. Two components of the decision context that decision makers should consider are risk distribution and the potential consequences of the decision.

Risk Distribution

Government decision makers face the challenge of acting on risk information for populations that may have widely varying exposures and sensitivities, and they must be careful to make sure that highly consequential risks are equitably distributed across these populations. In particular, if they consider only the average risks across an entire population, they might overlook the fact that there is a high risk in one group being offset by a low risk in another group. The social factors that can affect health risk estimates should have been identified at the problem-formulation stage and been analyzed during the assessment stage, and they should be presented to and considered by decision makers during this management phase of the decision-making process.

For example, a small number of individuals living near a petroleum refinery (the so-called maximally exposed individuals, or MEIs), might incur relatively large exposures to air pollutants, whereas much larger numbers of individuals might be exposed to significantly smaller levels (for example, the average level in an exposure distribution). The exposures between those average individuals and MEIs will often differ by more than a factor of 10. When making regulatory decisions, EPA should consider whether the primary basis for action is the protection of the smaller number of individuals incurring the larger risk or of the much larger number incurring smaller risks, and it should also consider how the magnitude of uncertainty in the MEI estimate relative to the uncertainty in the average exposure estimate should affect the basis of the decision.

Furthermore, it is likely that because of sex, genetics, life stage, nutritional status, occupational status (for example, pesticide applicators), or other factors, some individuals within a population will be more susceptible to adverse effects than the average for the general population. Looking again at the trichloroethylene (TCE) example discussed in Chapter 2 (see Box 2-4 for details), the range of potential human health endpoints (including carcinogenicity, developmental toxicity, and reproductive toxicity), potential exposure scenarios (including air, water, and soil), and variability within populations (due, for example, to sex, age, or nutritional status) generates complex and variable estimates of exposure, each with its own uncertainties, for different subpopulations. Those differing scenarios raise the question of which scenario or scenarios should be used to assess risks and, eventually, be used as the primary basis for regulatory action and also raises the issue of whether the uncertainties in the different scenarios should be taken into account. Although risk assessors develop the risk estimates, it is up to the regulatory decision makers to decide which estimates provide the most appropriate basis for setting standards. For example, if one wished to protect highly exposed individuals who would not be fully protected by a standard based on the lower estimates for the average population, one would choose to use exposure to the MEI population as the basis for the standard. Such a standard, however, would lead to more stringent regulatory controls and greater implementation costs than needed for the average population. By contrast, a standard geared to the average population might not provide adequate protection for the more highly exposed or more susceptible subpopulations. Decisions about which population is chosen to be the basis for setting the standard have both public health and economic consequences.

Potential Consequences of the Decision

Decisions that are made once and not reassessed are riskier than decisions that are revisited on a frequent, recurring basis. The ease with which a decision can be reversed or revisited at a later date or the degree to which a given decision precludes or enables additional choices at a later date will affect how uncertainty in risk estimates, cost–benefit analyses, technology assessments, and other factors are considered in the decision. The severity of the consequences of a decision—that is, whether the stakes of the decision are high or low—will also affect how that uncertainty is considered. For example, a decision that increases the likelihood of a nonsevere health outcome or that is not expensive (for example, with potential control costs that are small) will be easier to make in the face of uncertainty than a decision that increases the likelihood of a severe disease (such as cancer) or that will require expensive control technologies.

Using Uncertainty Analysis in Decision Making

Decision makers often want simple answers to questions like "Is this substance safe?" or "How can I set a regulatory standard that ensures absolute public safety?" Uncertainty analysis cannot provide unqualified answers to these questions. Instead, its results are usually stated in terms of ranges of numbers, likely or less likely outcomes, or probability distributions over health impacts. For example, an uncertainty analysis may result in an assessment that the most exposed individual has a very small risk, say 10^{-5}, of contracting cancer over his or her lifetime, with a possible uncertainty range of 10^{-4} to 10^{-6}. Another example may find that a population risk from exposure to fine particulates has a broad range, say, from 10 to 10,000 premature deaths, possibly further quantified by a probability distribution of outcomes over this range.

Because it lacks simple and unqualified answers, uncertainty analysis often complicates decision making. Nevertheless an honest expression of scientific uncertainty is an important part of any analysis to support decision making. Morgan and Henrion (1990) offer three reasons why it is important to perform an explicit treatment of uncertainty:

1. To identify important disagreements in a problem and to anticipate the unexpected;
2. To ask experts to be explicit about what they know and whether they really disagree in order to understand the experts and their often differing opinions; and
3. To update and adapt policy analyses that have been done in the past whenever new information becomes available, in order to improve decision making in the present.

Furthermore, it is important to provide responsible expressions of uncertainty to those making regulatory decisions because

1. Without understanding the uncertainty surrounding the key factors in a decision, decision makers may be tempted to use means or other central estimates and ignore unlikely, but extreme results.
2. Alternatively, without explicit expression of uncertainty, decision makers may be overly cautious and make decisions based on extremely conservative assumptions.

There are no simple rules for translating uncertainty information into a decision. However, a decision maker should be informed about and appreciate the range of uncertainty when making a decision. It is also helpful to present this information in a form that can assist decision making, such as by presenting probability distributions for health effects, examining extreme cases and tail probabilities, and incorporating these inputs into a more formal cost–benefit or decision analysis. Ultimately, decision makers have to make the decision in the face of uncertainty, a difficult job that involves weighing probabilities against consequences, protecting the average person as well as the most sensitive and exposed ones, and providing assurance that the regulatory action will be protective, even in the face of unlikely scenarios, model assumptions, and parameters.

The legal framework provides constraints on regulatory decisions (see the columns in Table 5-1), and how the uncertainty information is presented and used depends on the type of uncertainty (rows in Table 5-1). In the following sections we discuss the implications of uncertainty analysis for decision making for each of the three columns in Table 5-1, highlighting the differences between the three types of uncertainty (rows of Table 5-1) as appropriate.

Health Uncertainties

As discussed previously, under some environmental laws EPA is charged with protecting public health and not with balancing public health against other factors such as cost. In this case uncertainty analysis is necessarily restricted to assessing the uncertainty about health effects and the likely reduction of health effects for different regulatory options. In the case of variability and heterogeneity, uncertainty is usually expressed by presenting tables of risks for different populations and environmental conditions, together with a statistical assessment of the relative likelihood of the associated population and environmental categories. Often, those tables demonstrate the extreme cases, such as the most sensitive subpopulation or the highest-exposure conditions or both. Decision makers then have to make a difficult judgment concerning how to set an appropriate level of protection for the population at large and for those most sensitive or those exposed to the most severe environmental conditions. In many cases two specific hypothetical cases are examined: the average population risk and the risk to the maximally exposed or most sensitive person. The various possible regulatory options are then compared in terms of the risks for both cases.

When model and parameter uncertainty come into play, there is additional uncertainty about health risks. With model and parameter uncertainty, even the average risk to the population under average environmental conditions is subject to uncertainty. The decision maker has to make important judgments about how credible the extreme risk estimates are and how much to rely on them in decision making. It is important to compare risk estimates and associated distributions across the regulatory options and to characterize the risks with uncertainty (probability) distributions for all options. In many cases this will reveal a marginal decrease in risk reduction (both in terms of the mean risk and the high-risk tail) as the risk-reduction effort increases. The decision makers' task is then to weigh the marginal decrease in risk against the effort made to reduce it. When deep uncertainty is involved, tables of risks that characterize the risks for very different scenarios can help clarify the issues. For example, the Intergovernmental Panel on Climate Change (IPCC) has developed several reference scenarios that describe different future worlds based on assumptions about population growth, economic development, and patterns of production and consumption, especially in the energy area (IPCC, 2007). Those scenarios may involve both changes in the natural environment and major social and demographic changes. It is tempting to identify pessimistic scenarios (for example, a scenario in which climate change leads to higher levels of precipitation, population growth is larger than expected in some regions, and shifts to renewable and low-carbon forms of energy occur late). In the case of deep uncertainty, fair efforts should be made to identify a range of scenarios that identify all possible paths. Within each scenario, risk estimates can be provided for different regulatory options. That still leaves the decision makers with the difficult task of selecting regulatory options in the face of huge variations in scenarios. Decision makers should be informed by extensive sensitivity analyses and should examine the marginal decrease in risk reduction as the level of effort increases. The main goal is to find regulatory solutions that are effective over a broad spectrum of scenarios.

The above comments were aimed at the use of uncertainty analysis directly in decision making. As Morgan and Henrion (1990) have pointed out, there are many ancillary benefits to an explicit and formal treatment of uncertainty. Uncertainty analysis can demonstrate the difference in risks for different subpopulations and environmental conditions, thus providing a basis for the debate about trading off the average population versus sensitive or highly exposed ones. Uncertainty analysis is also useful in determining where experts agree and disagree, and it provides a focus for debate as well as for efforts to collect further information. Uncertainty analysis also helps stakeholders clarify how extreme values of the probability distributions affect regulatory decisions.

Uncertainties About Technology Availability

In examining uncertainties about technology availability case (column 2 of Table 5-1), we consider both health risks and the availability of technologies to reduce them, for example, by reducing air pollution from power plants or water pollution from chemical plants. Regulatory frameworks for this case often require the implementation of "best practicable" or "best available" technologies. Regarding the uncertainty analysis of health risks, the discussion above is still applicable. The new element is the uncertainty analysis of technological availability.

In assessing uncertainties about technology availability, the tools of technology assessment and technology risk analysis apply. Some technologies considered for implementation will be mature, proven, already in use, and immediately implementable at a reasonably well-known cost. Other technologies may only have been proven in principle and may have never been used for the purposes at hand. Assessing uncertainties about the likelihood of successfully developing the unproven technologies and the effectiveness of the technologies if they are successfully developed can inform the decision about which technology may be considered "best practicable" or "best available."

The results of an uncertainty analysis of technology availability can inform decision makers in quantitative terms about the maturity of the technology. This is almost always an issue involving expert judgment, and it is most likely to involve model and parameter uncertainty. Technology availability rarely involves variability or heterogeneity, and it involves deep uncertainty only for the most speculative technologies, for example, cold fusion.

Uncertainties About Cost–Benefit Analyses

In environmental regulatory contexts the benefits of regulations are reduced health and environmental risks. Health and environmental risks are uncertain and so are the benefits of reducing them with new regulations. The issue of using uncertainty analysis of health effects was discussed before. Here we discuss the use of uncertainty analysis of economic costs.

There are two types of economic costs. There are the direct life-cycle costs of implementing a proposed regulation through implementing new technologies and processes. These are often uncertain, especially with new technologies. There is also uncertainty about the broader economic impact of a proposed regulation, involving questions like, Will this decrease the competitiveness of Industry A? or, Will it cost jobs, and how many? In this section we will focus on direct economic costs.

Uncertainty analyses about direct costs can be used by decision makers in connection with uncertainty analysis about health risks and benefits to compare costs and benefits. These analyses will often show that both the health benefits and the costs are highly uncertain, and, as a result, decisions to change regulations are not easily differentiated. For example, in a graphical representation, with costs on the x axis and health impacts on the y axis, a representation without uncertainties would show different regulatory options as points in that graph; when considering uncertainties, these points would be surrounded both by vertical error bands (reflecting uncertainty about health effects) and by horizontal error bands (reflecting uncertainty about direct costs). Rarely do these representations suggest simple and straightforward solutions, but they still need to be presented to the decision maker to properly reflect the uncertainty inherent in the problem.

The uncertainty about macroeconomic impacts of proposed regulations (impact on special industries, gross domestic product, and employment) is often even more substantial. Here we find disagreement among experts about the way that regulation affects the economy, even when using similar models—for example, input–output models of computable generalized equilibrium models. Providing ranges and sensitivity analyses is the most common way to express uncertainties in these models.

Value of Information

In typical environmental and health problems, EPA decision makers must select among regulatory options. However, they also have the choice to defer a decision and to gather additional information prior to making a final decision. Value-of-information (VOI) methods can help determine whether or not additional information will help when making a decision.[3] VOI methods permit decision makers to compare alternative strategies for managing uncertainty: electing to proceed with currently available, uncertain information; electing to invest in better information that reduces the uncertainty prior to formulating a decision; or electing to ignore uncertainty entirely (NRC, 2009). In other words, VOI analysis "evaluates the benefit of collecting additional information to reduce or eliminate uncertainty in a specific decision making context" (Yokota and Thompson, 2004a, p. 635).

VOI analysis has been applied to business decision making (see Box 5-2) and medical decision making (Yokota and Thompson, 2004b).[4] Although not yet widely applied to environmental decisions (Yokota and Thompson, 2004a), the use of VOI has been recommended for such decisions (Presidential/Congressional Commission on Risk Assessment and Risk Management, 1997) and has been applied to some questions about climate change (see, for example, Nordhaus and Popp, 1997; Rabl and van der Zwaan, 2009; Yohe, 1996). Yokota et al. (2004) applied the VOI approach in the Voluntary Children's Chemical Evaluation Program initiated by EPA in 2000; working in the context of tiered chemical testing, they sought to answer the question of when information about the risks to children is sufficient.[5] Their analysis demonstrated that knowledge about exposure levels and control costs is important for decisions about toxicity tests.

As discussed by Hammitt and Cave (1991), the guiding principle of a VOI approach is that additional information is valued not for its own sake, but rather for the potential benefit of making better, welfare-enhancing decisions in the future. Findings from additional research are not known beforehand, so it is the expected value of the improvement in welfare that is relevant. For example, if finding A is obtained, we wish to know the decision and the utility (or gain), however measured, that is associated with the outcome, and the same for finding B, and so on.

Research that is unlikely to change a decision is considered to have little value ex ante. In some cases it may be possible to address uncertainty about a key underlying assumption, the "weakest link." If research can resolve

---

[3] Value-of-information analysis is one tool from the field of decision analysis. See Howard (2007) for a discussion of the field of decision analysis in general.

[4] Value-of-information analysis is referred to as expected value of information (EVI) analysis in medical decision making (Claxton et al., 2001).

[5] In the Voluntary Children's Chemical Evaluation Program (VCCEP), EPA asked the manufacturers or importers of 23 chemicals to which children have a high likelihood of exposure "to [voluntarily] provide information on health effects, exposure, risk, and data needs" (EPA, 2010).

> **BOX 5-2**
> **Implementing Value-of-Information Analysis in a Business Context**
>
> Suppose that a small division of a large company has been asked to maximize the expected profits of its division. The division has to make a decision now between two options for next year. A safe option yields $500,000 in profits next year, and a risky option yields returns that are contingent on a future event, which gives the firm $1 million if the event occurs and no profit if it does not. The event has a 40 percent chance of occurrence. Then the expected profit of the risky option is $400,000, and if there is no opportunity to collect more information, the firm should pursue the safe option.
>
> If the firm could collect "perfect information" in advance on whether the event will occur, it could take the risky option when the event is known to occur and the safe option when it is known not to occur, in which case the firm will realize an expected profit of $700,000 (0.40 × $1,000,000 + 0.60 × 500,000 = $700,000). Thus, the expected value of perfect information is $200,000 ($700,000, the profit with optimal decisions including the information, minus $500,000, the profit from the optimal decision without such information).
>
> Similarly, the "expected regret" could be calculated to indicate the value of information. The regret of choosing the safe option is zero if the event does not occur and $500,000 if it does (with $500,000 being the difference between the $1 million dollars the division you would get with the risky option and the $500,000 it is guaranteed with the safe option). Multiplying the regret value ($500,000) by the probability of the event (0.4) times regret yields $200,000 in expected regret, which is the same as calculated above from profit minus profit. The first fundamental theorem of VOI holds that expected regret equals VOI.

this uncertainty, which is critical to a choice among policy options, it may be particularly valuable.

The goal of a VOI analysis is to determine the value of additional information in coming to a decision. Although formal VOI analyses involve complex modeling, the essence of what is calculated in a VOI analysis can be explained in relatively simple terms. When calculating the VOI, it is helpful to consider four different possible approaches to decision making under uncertainty: (1) making a decision that takes into account perfect information regarding the state of nature; (2) making a decision that takes into account imperfect additional information regarding the state of nature; (3) making a decision that takes into account current uncertainty regarding the state of nature; and (4) making a decision that ignores uncertainty regarding the state of nature. As can be seen in Figure 5-2, each of these decision approaches can be placed along a horizontal axis that represents

168   ENVIRONMENTAL DECISIONS IN THE FACE OF UNCERTAINTY

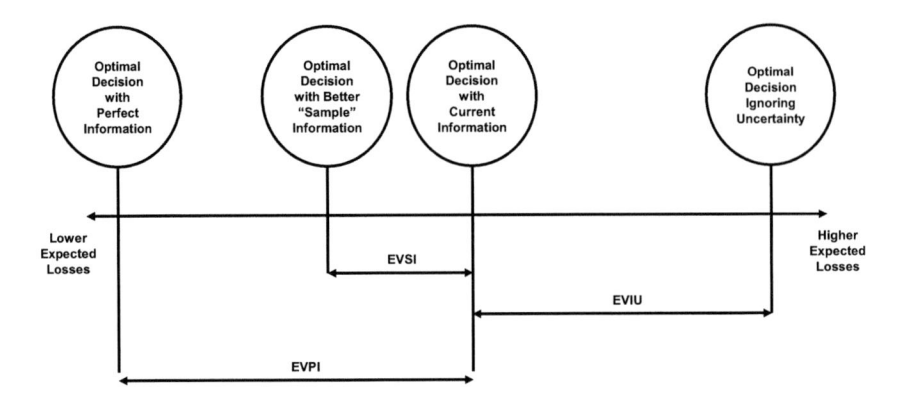

FIGURE 5-2 Schematic illustrating the values that can be calculated in a value-of-information analysis.
Abbreviations: EVIU = expected value of including uncertainty; EVPI = expected value of perfect information; EVSI = expected value of sample information.

the expected losses; the losses are least when acting optimally with perfect information and greatest when uncertainty is ignored. A number of different measures of the value of information can be calculated, including the expected value of perfect information (EVPI),[6] the expected value of sample information (EVSI), and the expected value of including uncertainty (EVIU) (see Box 5-3 for a description). The calculation of EVPI, EVSI, and EVIU is illustrated beneath the horizontal axis in Figure 5-2. All three measures compare the expected value of acting optimally with current information against the expected value of proceeding under reduced- or zero-uncertainty conditions. EVPI captures the value of *eliminating* uncertainty; EVSI captures the value of *reducing* uncertainty; and EVIU captures the value of *ignoring* uncertainty.

A value-of-information analysis has a number of benefits. First, it captures the sensitivity of decisions to uncertainty, taking into explicit account the decision maker's level of risk aversion, the inherent variability of the situation, and the current state of the evidence base. Second, it can serve as a guide to model selection and justification. In instances where EVIU greatly exceeds EVPI, for example, analysts may find it easier to proceed with current evidence, knowing that the inclusion of uncertainty in their models is what matters and that delaying the analysis in anticipation of

---

[6] A more precise term for the expected value of perfect information (EVPI) is the maximum amount a person or society is willing to pay for the information, which incorporates attitudes toward risk.

> **BOX 5-3**
> **Examples of Value-of-Information Measures**
>
> **The expected value of including uncertainty (EVIU)**
> The EVIU is defined as the improvement in net benefit (or avoided harm) that can be achieved when the uncertainty surrounding a decision is taken into account. The EVIU is computed by taking the difference between the expected outcome that could be achieved by making optimal use of currently available information about uncertainty and the expected outcome that could be achieved by ignoring that uncertainty and simply treating all random variables as fixed at some central, point estimate value. The resultant figure provides an upper bound on the value of building uncertainty into the analysis in the first place.
>
> **The expected value of perfect information (EVPI)**
> EVPI is the improvement in net benefit (or avoided harm) that can be achieved if the uncertainty surrounding a decision is completely resolved. The EVPI is computed by taking the difference between the expected outcome that would result from making optimal use of perfect information and the expected outcome that would result from making optimal use of currently available information. The resultant figure provides an upper bound on the value of additional investment in information. It denotes the most a decision maker should be prepared give up to learn the true state of nature. Although an EVPI analysis postulates the impossible situation in which all variability is eliminated, it offers decision makers useful insights into the extent to which current uncertainty reduces the quality of their decisions.
>
> **The expected value of sample information (EVSI)**
> The EVSI is closely related to the EVPI. Recognizing that it is almost never possible to completely eliminate uncertainty, the EVSI measures the improvement in net benefit (or avoided harm) that could be achieved if the uncertainty surrounding a decision could be reduced, rather than completely resolved. Here the expected outcome that could be achieved by making optimal use of currently available information is compared with the expected outcome that could be achieved by making optimal use of some specified level of additional information.

better information will confer little additional value. By contrast, in instances where EVIU is comparatively small, analysts may be more able to justify using fixed point estimates and ignoring parameter uncertainty. Third, high ratios of EVSI to EVPI may indicate the presence of efficient research investment opportunities, thus helping decision makers to identify priorities from among competing sources of uncertainty.

The decision to gather more information in response to a VOI analysis can delay the actual decision, and that delay has costs associated with it. Those costs include (1) the costs of additional data collection and analysis;

(2) in many situations, more importantly, the cost of delaying action; and (3) the cost of modifying decisions once implemented. Some decisions are, for all practical purposes, irreversible. If the costs of data collection or research and delaying a decision are low and the costs of subsequently modifying a policy decision are high, decision makers may decide to seek further information to reduce uncertainty before making a decision.

Business decisions—where the value of information is the difference between the profits with and without the information—illustrate well the concept of VOI analysis (see Box 5-2 for an example). As can be seen in Box 5-2, the value of information is not a fixed number but rather a random variable that depends on the decision maker's prior estimate of what the new information will reveal. For this reason, the term "expected value of information" is used, referring to what the additional information is expected to be worth on average before the new information is collected. If there is no possible outcome in which additional information gathering or research would change the decision, then the expected value of information is zero. If any decision is changed for the better after some result, than the value of information is positive. If the costs of obtaining the information (either data-gathering research costs or costs from delaying a decision, as might be the case with some regulatory options) are less than the expected value of information, then it is better to get the information.[7]

In the theoretical world of Box 5-2, it is assumed that the additional information is perfect. For example, there are no errors in predicting whether or not an event will occur. In practice, however, errors are made. The prediction may fail to predict that an event will or will not occur. In the real world of imperfect information, one can use Bayesian updating to incorporate the uncertainty inherent in the new information. In Bayesian updating, a weight is attached to new information, and a second weight is attached to the prior belief; the weights must sum to one. Thus, if the new information is thought to be particularly credible, it will be assigned a higher weight, with a correspondingly lower weight being placed on the prior belief. In the context of VOI, if the weight on placed on the new estimate is low, then it will generally not pay to obtain the additional information (Hunink, 2001; Raiffa, 1968).

In a public context, such as EPA's decision context, the value of information is calculated in terms of the anticipated net benefits rather than the anticipated profitability calculated in the business example. The calculation of net benefits can be very complex, and for many decisions the only

---

[7] Conventionally the costs of obtaining the information are not part of the value-of-information calculations, but are instead compared at the end. In any case, these costs and the cost of revising the decision once made must be considered in choosing whether or not to seek additional information and, if so, what type of information is to be sought.

practical way to assess the effect of a policy option, as well as its inherent uncertainty, is to implement the option and pursue an active strategy of monitoring its effects, with the monitoring being done, to the extent possible, in quantitative terms.

While conceptually appealing, in the context of EPA's public decision making VOI has two challenges that do not create as many difficulties in private decisions, such as the one described in Box 5-2. First, expected profit is simpler to estimate than the estimated costs involved in a decision by the EPA. Second, unlike the situation with private decisions, the rationale for EPA postponing a decision and seeking new information may have to be explained to various segments of the public. That explanation will be complicated by the fact that many of the costs and assumptions underlying VOI calculations, including the credibility weights in Bayesian updating analysis, are subjective and difficult to defend. Despite those challenges, VOI can be a useful approach to help determine what information is worth gathering for future decisions.

Decision Implementation

Implementation of a regulatory decision is an important step in the management process. This step requires significant skills in addressing often competing stakeholder, legal, and political considerations surrounding the proposed decision.

Good decision making under uncertainty involves updating information through research, monitoring the implementation of regulatory action, and periodically revisiting and adapting the decision. A plan should be in place that outlines which uncertainties are being researched and when the decision will be revisited to see if uncertainty has decreased to the point that the decision should be revisited. As discussed earlier in this chapter, when decisions involve deep uncertainty, adaptive management approaches are particularly useful. Those approaches require increased monitoring and a plan for gathering more information and revisiting the decision.

OTHER CONSIDERATIONS

As discussed in Chapter 1, other factors in addition to human health risks, economic factors, and technology availability play an important role in many of EPA's decisions. Although not thought of as traditional uncertainties that can be quantified, there is uncertainty in those factors that should be considered in making and communicating about with EPA's decisions. The roles that some of those factors and the uncertainties in them play in EPA's decisions are discussed below.

Special Populations and Equity

In some cases the regulatory problem is shaped by issues concerning special populations (e.g., the lead exposure of children) or by equity or environmental justice concerns, which have been labeled as priorities by various executive orders,[8] although these orders do not have the weight of law. EPA recently issued a report, *Plan EJ 2014: Legal Tools*, that details the legal tools related to environmental justice that are available to the agency (EPA, 2011b).

These special considerations could influence the choice of analytical approaches since approaches that emphasize net aggregate costs and benefits do not typically address these concerns. In particular, these factors can add to the variability and heterogeneity in estimates of health risks and economic factors. If the formal approaches described above are used in these contexts, they must be disaggregated so that the impacts they have on special populations can be examined as well as the aggregate effects. In doing so, EPA will be able to see the effects that its decisions could have on different groups and will be able to include the potential effects on those groups in the rationale for its decision. That will allow stakeholders to better understand the agency's decision.

Geographic Scope

The geographic scope of a decision problem may be global, national, regional, or local. Spatial or geographic considerations are likely to introduce special problems into assessing and accounting for uncertainty. For example, data on a local area may be inadequate to characterize exposure or the sensitivity of populations to the exposure. Given the inadequacy of data collected on a national basis for use in decisions limited to local areas, decision making may be improved by additional data collection and analysis. Furthermore, the preferences of the residents in a community may differ from national averages, and those preferences can affect the values that people assign to outcomes which, in turn, will affect the economic analyses. The goal of an uncertainty analysis is to characterize how these values differ, and doing so may require additional data collection. A characterization of such differences can be qualitative or quantitative.

If the scope of a problem is local, such as is the case for a Superfund problem, local stakeholders (including members of the public) may provide input at various times during the analysis phase. It is crucial, therefore, to obtain stakeholder involvement in the problem-formulation phase,

---

[8] For example, Exec. Order No. 12898. 77 FR 11752 (February 28, 2012) and Exec. Order No. 13045. 78 FR 19884 (April 23, 1997).

particularly with regard to decisions about the endpoints to be included in the analyses. On the other hand, if the scope of a problem is national, as is the case when setting an ambient air quality standard, the type of stakeholder involvement will be driven more by the statutory framework and agency procedures. For national issues, the stakeholders who provide input are often representatives of groups with special interests (e.g., industry, or advocacy organizations focused on a particular disease) in addition to—or even rather than—being community members.

Decisions applicable to a specific geographic area are well suited to the incorporation of public values. Even when the statutory directive is for the consideration of health effects, the implementation plans will often be of great interest to local communities. For this reason EPA will often solicit input on implementation plans through written comments or at hearings in order to gather public comments at locations across the country (EPA, 2012).

Identifying the effects of geographic scope on a decision in the initial, problem-formulation stage will help EPA identify important stakeholders and ensure that the variability in the perspectives can be addressed in the assessment and management phases of the decision. These concerns could affect the assessment of economic factors in particular.

Uncertainty analysis and more formal approaches to decision making have not always been applied to these factors in a systematic or rigorous way, but some of the analytic techniques described in Chapter 2 and Appendix A could be applied to them. For example, Arvai and Gregory (2003) used multiattribute utility analysis to evaluate different approaches to stakeholder involvement in a decision related to the cleanup of a contaminated site; one approach involved the presentation of scientific information, while the other involved the presentation of scientific information and "values-oriented information that seeks to improve the ability of nonexpert participants to make difficult trade-offs across a variety of technical and nontechnical concerns" (p. 1470). The importance of stakeholder engagement is discussed further below.

STAKEHOLDER ENGAGEMENT

Agency decision-making processes that involve stakeholders, including dialogues with stakeholders about uncertainties, can demonstrate intentional transparency and create, maintain, and enhance a relationship of trust between the agency and its stakeholders.[9] In addition, a growing

---

[9] The terms used to refer to the parties that can be involved in environmental decision making are varied and include "stakeholders," "the public," "affected parties," and "interested parties." The definitions of these terms (i.e., the expertise, affiliations, and perspectives of the

body of research demonstrates that the political aspects of stakeholder processes do not sacrifice decision quality (Beierle, 2000) and that public participation (NRC, 2008) can in fact add information to and improve the quality and legitimacy of agencies' decisions about the environment.[10] Because decisions may ultimately have some impact for the stakeholders, if the decision-making process is to be fair and democratic stakeholders must be given the opportunity to be involved in making those decisions, including decisions about which uncertainties need better elucidation. Early and continuous involvement of stakeholders can also prevent delays that can occur when stakeholders are not engaged in decision making until later in the process, at which time they might take legal actions.

EPA has issued much guidance on public and stakeholder involvement in its programs and activities (EPA, 1998, 2003, 2011a), and there are several regulations that contain public involvement procedures for specific EPA programs and activities.[11] The EPA also issued an agency-wide public involvement policy (reissued periodically with updates) that can be applied to all EPA programs and activities (EPA, 2003).[12] The agency-wide policy is not mandatory, however. In spite of the existing guidance, there has been repeated concern and criticism over the failure of EPA to engage stakeholders more systematically and adequately as part of its various regulatory mandates for environmental decision making (see, for example, NRC, 1996, 2008; Presidential/Congressional Commission on Risk Assessment and Risk Management, 1997). This was the justification for a recommendation made in *Science and Decisions* (NRC, 2009) that EPA adopt formal provisions for stakeholder involvement across a three-phase framework for

---

individuals and organizations they include) have also varied. Unless otherwise specified, in this report we use *stakeholder* to refer to any parties interested in or affected by a decision-making authority's activities. Stakeholders may include decision makers, industry groups, communities and community organizations, environmental organizations, scientists and technical specialists, individuals from the public, and others.

[10] For a comprehensive review of research on public participation in environmental assessment and decision making, the reader is encouraged to refer to NRC, 2008.

[11] See, for example, 40 CFR Part 25—Public Participation in Programs under the Resource Conservation and Recovery Act, the Safe Drinking Water Act, and the Clean Water Act; 40 CFR Part 271—Requirements for Authorization of State Hazardous Waste Programs; 40 CFR Part 300—National Oil and Hazardous Substances Pollution Contingency Plan, Subpart E—Hazardous Substance Response (establishes methods and criteria for determining the appropriate extent of response authorized by CERCLA and CWA section 311(c)).

[12] According to the guidance, the seven basic steps to effective public involvement are to (1) plan and budget for public involvement activities, (2) identify the interested and affected public, (3) consider providing technical or financial assistance to the public to facilitate involvement, (4) provide information and outreach to the public, (5) conduct public consultation and involvement activities, (6) review and use input and provide feedback to the public, and (7) evaluate public involvement activities (EPA, 2003).

risk-based decision making (see Figure 5-1).[13] This recommendation echoes the point made in other NRC reports (see, for example, NRC, 1996, 2008) that technical and analytical aspects of the decision-making process be balanced with adequate involvement by interested and affected parties, and it is a point with which this committee concurs.

Concerns about procedural fairness and trust are even more salient when scientific uncertainty is reported (NRC, 2008). Some research has demonstrated that people show a heightened interest in evaluating the credibility of information sources when they perceive uncertainty (Brashers, 2001; Halfacre et al., 2000; van den Bos, 2001), and they are also more likely to challenge the reliability and adequacy of risk estimates and be less accepting of reassurances in such situations (Kroll-Smith and Couch, 1991; Rich et al., 1995). When EPA anticipates more uncertainty in scientific aspects of decision making, the need for stakeholder involvement may often be greater. Other research has spoken to the importance of describing the existence of uncertainties in risk assessments as well, both to facilitate transparency and to increase public perceptions of agency honesty (Johnson and Slovic, 1995; Lundgren and McMakin, 2004; Morgan and Henrion, 1990; NRC, 1989).

Developing provisions for stakeholder involvement in decision making, including guidance on discussing with stakeholders the sources of uncertainty and how uncertainty is being managed, could lead to greater transparency and trust and also has the potential to result in better decision making. Stakeholders might be interested in how uncertainty can be dealt with in the analysis, the implications of uncertainties, and what can or cannot be done about the uncertainties. Stakeholders may also suggest new uncertainties not previously under consideration by EPA and, by expressing their values and concerns (cultural, religious, economic, and so on), help decision makers prioritize how the uncertainties are factored into decision making.

In discussions with stakeholders about uncertainty, it is important that EPA be proactive in engaging the range of stakeholders for whom a decision may have an impact. *Science and Decisions* (NRC, 2009) recommended that EPA provide incentives to allow for balanced participation of stakeholders, including affected communities and those stakeholders for whom participation is less likely because of competing priorities, fewer resources, a lack of knowledge, or other factors. Boeckmann and

---

[13] The three phases are (1) problem formulation and scoping, (2) planning and conduct of risk assessment, and (3) risk-management phases (see Figure 5-1). As part of the framework, the report also suggests that stakeholder involvement should have time limits so as not to delay decision making and that there should be incentives so that participation is more balanced and includes impacted communities and less advantaged stakeholders (NRC, 2008).

Tyler (2002) found that the public is more likely to participate "in their communities when they feel that they are respected members of those communities" (p. 2067). Showing respect, therefore, is important for stakeholder engagement. The resources required for such an engagement of stakeholders, however, must be weighed against the need for such actions, given the context of the decision, including consideration of the potential health risks, the costs associated with the potential regulatory options, and the magnitude, sources, and nature or type of the uncertainty associated with the decision.

KEY FINDINGS

- Incorporating uncertainty analysis into a systematic framework, such as a modified version of the decision framework in *Science and Decisions* (NRC, 2009), provides a process for decision makers, stakeholders, and analysts to discuss the appropriate and necessary uncertainty analyses.
- Involvement of decision makers in the planning and scoping of uncertainty analyses during the initial, problem-formulation phase will help ensure that the goals of the uncertainty analysis are consistent with the needs of the decision makers.
- Involvement of stakeholders in the planning and scoping of uncertainty analyses during the initial problem-formulation phase will help define analytic endpoints and identify population subgroups as well as heterogeneity and other uncertainties.
- Uncertainty analysis must be designed on a case-by-case basis. The choice of uncertainty analysis depends on the context of the decision, including the nature or type of uncertainty (that is, heterogeneity and variability, model and parameter uncertainty, or deep uncertainty), and the factors that are considered in the decision (that is, health risk, technology availability, and economic, social, and political factors) as well as the data that are available.
- When assessing variability and heterogeneity:
  o Analyses of statistical distributions, including extreme-value analyses, are useful for assessing uncertainty in data on health effects (that is, estimates of risks). The use of safety or default factors (using statistics) can also be helpful under certain circumstances.
  o Direct assessments and technological choice or risk analyses developed using statistics can be helpful for assessing technological availability.

- Cost-effectiveness, cost–benefit analysis, and multiattribute utility analysis developed using statistical methods can be useful for assessing costs and benefits.
- When assessing model and parameter uncertainty:
  - Expert elicitation and the analysis of probability distributions, including extreme value analyses, can be useful for assessing health effects. Safety or default factors developed using expert judgments can also be helpful.
  - Formal expert elicitation to assess technology availability, as well as technology choice and risk analysis using expert judgment, can be helpful in assessing technology factors.
  - Cost-effectiveness, cost–benefit analysis, and multiattribute utility analysis developed using expert judgments can be useful for assessing costs and benefits.
- When assessing deep uncertainty:
  - Scenario analysis and robust decision-making methods can be helpful for assessing health effects, technology factors, and costs and benefits.
- The interpretation and incorporation of uncertainty into environmental decisions will depend on a number of characteristics of the risks and the decision. Those characteristics include the distribution of the risks, the decision makers' risk aversion, and the potential consequences of the decision.
- The quality of the analysis and recommendations following from the analysis will depend on the relationship between analyst and the decision maker. The planning, conduct, and results of uncertainty analysis should not be isolated from the individuals who will eventually make the decisions. The success of a decision in the face of uncertainty depends on the analysts having a good understanding of the context of the decision and the information needed by the decision makers, and the decision makers having a good understanding of the evidence on which the decision is based, including an understanding of the uncertainty in that evidence.

RECOMMENDATION 7

Although some analysis and description of uncertainty is always important, how many and what types of uncertainty analyses are carried out should depend on the specific decision problem at hand. The effort to analyze specific uncertainties through probabilistic risk assessment or quantitative uncertainty analysis should be guided by the ability of those analyses to affect the environmental decision.

REFERENCES

Arvai, J., and R. Gregory. 2003. Testing alternative decision approaches for identifying cleanup priorities at contaminated sites. *Environmental Science and Technology* 37(8):1469–1476.
Beierle, T. C. 2000. *The quality of stakeholder-based decisions: Lessons from the case study record.* Washington, DC: Resources for the Future.
Boeckmann, R. J., and T. R. Tyler. 2002. Trust, respect, and the psychology of political engagement. *Journal of Applied Social Psychology* 32(10):2067–2088.
Brashers, D. E. 2001. Communication and uncertainty management. *Journal of Communication* 51(3):477–497.
Claxton, K., P. Neumann, S. Araki, and M. Weinstein. 2001. Bayesian value-of-information analysis. An application to a policy model of Alzheimer's disease. *International Journal of Technology Assessment in Health Care* 17(1):38–55.
EPA (U.S. Environmental Protection Agency). 1998. EPA stakeholder involvement action plan. Washington, DC: Environmental Protection Agency. http://www.epa.gov/publicinvolvement/siap1298.htm (accessed November 20, 2012).
———. 2003. *Public involvement policy of the U.S. Environmental Protection Agency.* Washington, DC: Environmental Protection Agency. http://www.epa.gov/publicinvolvement/pdf/policy2003.pdf (accessed January 3, 2013).
———. 2010. Voluntary Children's Chemical Evaluation Program (VCCEP). http://www.epa.gov/oppt/vccep (accessed November 20, 2012).
———. 2011a. *Expert Elicitation Task Force white paper.* Washington, DC: Environmental Protection Agency. http://www.epa.gov/stpc/pdfs/ee-white-paper-final.pdf (accessed January 3, 2013).
———. 2011b. Plan EJ 2014: Legal tools. Washington, DC: Environmental Protection Agency. http://www.epa.gov/environmentaljustice/resources/policy/plan-ej-2014/ej-legal-tools.pdf (accessed January 3, 2013).
———. 2012. *The plain English guide to the Clean Air Act.* http://www.epa.gov/air/caa/peg/public.html (accessed May 24, 2012).
Fenwick, E., K. Claxton, and M. Sculpher. 2001. Representing uncertainty: The role of cost-effectiveness acceptability curves. *Health Economics* 10(8):779–787.
Garber, A. M., and C. E. Phelps. 1992. *Economic foundations of cost-effective analysis.* NBER working paper series no. 4164. Cambridge, MA: National Bureau of Economic Research. http://www.nber.org/papers/w4164.pdf (accessed January 3, 2013).
Gregory, R. 2011. *Structured decision making: A practical guide to environmental management choices.* Hoboken, NJ: Wiley-Blackwell.
Gregory, R. S., and R. L. Keeney. 2002. Making smarter environmental management decisions. *Journal of the American Water Resources Association* 38:1601–1612.
Gregory, R., T. C. Brown, and J. L. Knetsch. 1996. Valuing risks to the environment. *Annals of the American Academy of Political and Social Science* 545:54–63.
Halfacre, A. C., A. R. Matheny, and W. A. Rosenbaum. 2000. Regulating contested local hazards: Is constructive dialogue possible among participants in community risk management? *Policy Studies Journal* 28(3):648–667.
Hammitt, J. K., and J. A. K. Cave. 1991. *Research planning for food safety: A value of information approach.* Washington, DC: RAND.
Hammond, K. R., B. F. Anderson, J. Sutherland, and B. Marvin. 1984. Improving scientists' judgments of risk. *Risk Analysis* 4(1):69–78.
Howard, R. A. 2007. The foundations of decision analysis revisited. In *Advances in decision analysis*, edited by W. Edwards, R. F. Miles, and D. Von Winterfeldt. New York: Cambridge University Press. Pp. 32–56.

Hunink, M. G. M. 2001. *Decision making in health and medicine: Integrating evidence and values*. Cambridge; New York: Cambridge University Press.
IPCC (International Panel on Climate Change). 2007. *Climate change 2007: Synthesis report. Contribution of Working Group I, II and III to the Fourth Assessment Report of the Intergovernmental Panel on Climate Change*. http://www.ipcc.ch/publications_and_data/publications_ipcc_fourth_assessment_report_synthesis_report.htm (accessed November 12, 2012).
Johnson, B. B., and P. Slovic. 1995. Presenting uncertainty in health risk assessment: Initial studies of its effects on risk perception and trust. *Risk Analysis* 15(4):485–494.
Keeney, R. L. 1996. *Value-focused thinking: A path to creative decision making*. Cambridge, MA: Harvard University Press.
Keeney, R. L., and H. Raiffa. 1976. *Decisions with multiple objectives*. Cambridge, UK: Cambridge University Press.
Koppenjan, J., and E.-H. Klijn. 2004. *Managing uncertainties in networks: A network approach to problem solving and decision making*. London: Routledge.
Kroll-Smith, J. S., and S. R. Couch. 1991. As if exposure to toxins were not enough: The social and cultural system as a secondary stressor. *Environmental Health Perspectives* 95:61–66.
Lundgren, R. E., and A. H. McMakin. 2004. *Risk communication: A handbook for communicating environmental, safety, and health risks*. Columbus: Battelle Press.
Mingers, J., and J. Rosenhead. 2004. Problem structuring methods in action. *European Journal of Operational Research* 152(3):530–554.
Morgan, M. G., and M. Henrion. 1990. *Uncertainty: A guide to dealing with uncertainty in quantitative risk and policy analysis*. New York: Cambridge University Press.
Nordhaus, W. D., and D. Popp. 1997. What is the value of scientific knowledge? An application to global warming using the price model. *Energy Journal* 18(1):1–45.
NRC (National Research Council). 1983. *Risk assessment in the federal government: Managing the process*. Washington, DC: National Academy Press.
———. 1989. *Improving risk communication*. Washington, DC: National Academy Press.
———. 1996. *Understanding risk: Informing decisions in a democratic society*. Washington, DC: National Academy Press.
———. 2008. *Public participation in environmental assessment and decision making*. Washington, DC: The National Academies Press.
———. 2009. *Science and decisions: Advancing risk assessment*. Washington, DC: The National Academies Press.
Presidential/Congressional Commission on Risk Assessment and Risk Management. 1997. *Risk assessment and risk management in regulatory decision-making. Final report. Volume 2*. Washington, DC: Presidential/Congressional Commission on Risk Assessment and Risk Management.
Rabl, A., and B. van der Zwaan. 2009. Cost–benefit analysis of climate change dynamics: Uncertainties and the value of information. *Climatic Change* 96(3):313–333.
Raiffa, H. 1968. *Decision analysis: Introductory lectures on choices under uncertainty*. Reading, MA: Addison-Wesley.
Rich, R. C., M. Edelstein, W. K. Hallman, and A. H. Wandersman. 1995. Citizen participation and empowerment: The case of local environmental hazards. *American Journal of Community Psychology* 23(5):657–676.
Rosenhead, J. 1996. What's the problem? An introduction to problem structuring methods. *Interfaces* 117–131.
Spetzler, C. S. 2007. Building decision competency in organizations. In *Advances in decision analysis: From foundations to applications*, edited by W. Edwards, R. F. Miles, and D. von Winterfeldt. New York: Cambridge University Press. Pp. 451–468.

Stinnett, A. A., and J. Mullahy. 1998. Net health benefits: A new framework for the analysis of uncertainty in cost-effectiveness analysis. *Medical Decision Making* 18(2 Suppl.):S68–S80.
van den Bos, K. 2001. Uncertainty management: The influence of uncertainty salience on reactions to perceived procedural fairness. *Journal of Personality and Social Psychology* 80(6):931–941.
von Winterfeldt, D. 2007. Defining a decision analytic structure. In *Advances in decision analysis: From foundations to applications*, edited by W. Edwards, R. Miles, and D. von Winterfeldt. New York: Cambridge University Press. Pp. 81–103.
von Winterfeldt, D., and B. Fasolo. 2009. Structuring decision problems: A case study and reflections for practitioners. *European Journal of Operational Research* 199(3):857–866.
Yohe, G. 1996. Exercises in hedging against extreme consequences of global change and the expected value of information. *Global Environmental Change* 6(2):87–101.
Yokota, F., and K. M. Thompson. 2004a. Value of information analysis in environmental health risk management decisions: Past, present, and future. *Risk Analysis* 24(3):635–650.
———. 2004b. Value of information literature analysis: A review of applications in health risk management. *Medical Decision Making* 24(3):287–298.
Yokota, F., G. Gray, J. K. Hammitt, and K. M. Thompson. 2004. Tiered chemical testing: A value of information approach. *Risk Analysis* 24(6):1625–1639.

6

Communication of Uncertainty

Communication of uncertainty is an important component of the broader practice of human health risk communication. As discussed by Stirling (2010), conveying the uncertainty in the science related to the decision is crucial not only so that decision makers will understand the range of evidence on which to base a decision, but also because it can make the influences of "deep intractabilities of uncertainty, the perils of group dynamics or the perturbing effect of power . . . more rigorously explicit and democratically accountable" (p. 1031).

The U.S. Environmental Protection Agency (EPA) requested guidance on communicating uncertainty to ensure the appropriate use of risk information and to enhance the understanding of uncertainty among the users of risk information, such as risk managers (that is, decision makers), journalists, and citizens. Although, as discussed in the previous chapters, a number of factors play a role in EPA's decisions, most of the research the committee identified on the communication of environmental decisions focuses on communication of the uncertainty in estimates of human health risk, and those uncertainties are the focus of this chapter.

This chapter begins with background information on the communication of those risks. It then discusses the advantages and disadvantages of different formats for the presentation of uncertainty and the considerations that go into determining a communication strategy, such as the purpose of the communication, the stage in the decision-making process, the decision context, the type of the uncertainty, and the characteristics of the audience. The relevant audience characteristics discussed here are the audience's level of technical expertise, personal and group biases, and social trust. In response to its charge, when discussing audience characteristics, the committee paid special attention to communicating with the media.

COMMUNICATION OF UNCERTAINTY IN RISK ESTIMATES

The science of risk communication and the idea of what is good and appropriate risk communication have evolved over the past decades (Fischhoff, 1995; Leiss, 1996). For example, Fischhoff (1995) described the first three stages in that evolution in terms of how communicators think about the process: "All we have to do is get the numbers right," "All we have to do is tell them the numbers," and "All we have to do is explain what we mean by the numbers" (p. 138). With the realization that those factors alone would not lead stakeholders to accept decisions about risks, Fischhoff writes, communication experts changed strategies to include the ideas that "All we have to do is show them that they've accepted similar risks in the past," "All we have to do is show them that it's a good deal for them," and "All we have to do is treat them nice" (p. 138). Those approaches eventually evolved to include the current strategy, "All we have to do is make them partners" (p. 138). That most recent strategy includes both the two-way communication and the stakeholder engagement that today are considered hallmarks of good risk communication. In the context of EPA's decisions, stakeholders include the decision makers at the agency, the industries potentially affected by a regulatory decision, and individuals or groups affected by the decision, including local community members for local issues or all the public for issues of national significance.

Improving Risk Communication (NRC, 1989) emphasized the importance of such two-way communication for agencies such as EPA, defining risk communication as "an interactive process of exchange of information and opinion among individuals, groups, and institutions. It involves multiple messages about the nature of risk and other messages, not strictly about risk, that express concerns, opinions, or reactions to risk messages or to legal and institutional arrangements for risk management" (p. 21). That report noted that risk estimates always have inherent uncertainties and that scientists often disagree about the appropriate estimates of risk. It recommended not minimizing the existence of uncertainty, disclosing scientific disagreements, and communicating "some indication of the level of confidence of estimates and the significance of scientific uncertainty" (p. 170). Other reports have also emphasized the need for engagement of stakeholders in the decision-making process (NRC, 1996, 2009).

Documenting the type and magnitude of uncertainty in a decision is not only important at the time of the decision, as discussed by Bazelon (1974), but it is also important when a decision might be revisited or evaluated in the future.

The committee agrees with many of those concepts discussed and recommended in *Improving Risk Communication* (NRC, 1989). Many of those concepts have been incorporated into EPA guidance documents on risk communication (see, for example, Covello and Allen, 1988; EPA, 2004, 2007). In 2007, for example, EPA's Office of Research and Development published *Risk Communication in Action* (EPA, 2007), which describes the basic concepts of successful risk communication, taking into account differences in values and risk perception, and includes instructions on how best to engage with and present risk information to the public. It is not clear, however, the extent to which that and other documents—which are not agency-wide policies—are considered by or implemented in the risk communication practices of different programs and offices at EPA. Other National Research Council (NRC) reports (1996, 2008) have expressed concern that stakeholders have not been adequately involved in EPA decision making, suggesting that two-way risk communication, including communication surrounding uncertainty, may in some instances be inadequate.

The extent to which uncertainty is described and discussed varies among EPA's decision documents. Chapter 2 discusses EPA's decisions and supporting documentation around arsenic in drinking water, the Clean Air Interstate Rule (CAIR), and methylmercury, including the uncertainty analyses that EPA conducted and presented for those regulatory decisions. Those examples indicate that EPA does sometimes conduct numerous uncertainty analyses and present those analyses in its documents. Such analyses, however, are often presented in appendixes, and the ranges of potential outcomes are not necessarily presented in the summaries and summary tables. The committee also noted that the uncertainty analyses in those documents focus almost exclusively on the uncertainty in estimates related to human health risks and benefits. Krupnick et al. (2006) reviewed four of EPA's regulatory impact analyses for air pollution regulations, including CAIR and the Clean Air Mercury Rule. They concluded that although the documents "indicate increased use of uncertainty analysis," the EPA's regulatory impact analyses "do not adequately represent uncertainties around 'best estimates,' do not incorporate uncertainties into primary analyses, include limited uncertainty and sensitivity analyses, and make little attempt to present the results of these analyses in comprehensive way" (p. 7).

To successfully communicate uncertainty, EPA programs and offices need to develop communication plans that include identification of stakeholder values, perceptions, concerns, and information needs related to uncertainty about the decisions to be made and to the uncertainties to be evaluated. As discussed in Chapter 5, the development of those plans should be initiated in the problem-formulation phase of decision making, and it should continue during the assessment and management phases.

PRESENTATION OF UNCERTAINTY

The most widely used formal language of uncertainty in risk estimates is probability[1] (Morgan, 2009). As Spiegelhalter et al. (2011) stated, however, "probabilities are notoriously difficult to communicate effectively to lay audiences." Probabilistic information, and the uncertainties associated with those probabilities, can be communicated using numeric, verbal, or graphic formats, and consideration should be given to which approach is most appropriate. In a recent review Spiegelhalter et al. (2011) pointed out that the available research in this area for the most part is limited to small studies, often on students or self-selected samples. That lack of large, randomized experiments remains years after Bostrom and Löfstedt (2003) "concluded that risk communication was 'still more art than science, relying as it often does in practice on good intuition rather than well-researched principles'" (Spiegelhalter et al., 2011, p. 1399).

As discussed later in this chapter, the most appropriate approach to communicating uncertainty depends on the circumstances (Fagerlin et al., 2007; Nelson et al., 2009; Spiegelhalter et al., 2011; Visschers et al., 2009). Lipkus (2007) summarized the general strengths and weaknesses of each of the different approaches for conveying probabilistic information, based on a comprehensive literature review and consultation with risk communication experts (see Box 6-1). The committee discusses relevant findings from this research below. Regardless of the format in which the uncertainty is presented, it is important to bound the uncertainty and to describe the effect it might have on a regulatory decision. Presenting the results of analyses such as the sensitivity analyses and scenarios discussed in Chapter 5 is one way to provide some boundaries on the effects of those uncertainties and to educate stakeholders about how those uncertainties might affect a decision. It is important to note that the existence of weaknesses does not necessarily indicate that a given method should not be used, but rather those weaknesses should be considered and adjusted for when developing a communication strategy.

Numeric Presentation of Uncertainty

In general, numeric presentations of probabilistic information—such as presenting information in terms of percentages and frequencies—can lead to more accurate perceptions of risk than verbal and graphic formats (Budescu et al., 2009). Unlike graphic and verbal presentations, numeric information can be put into tables in order to communicate a large amount of information in a single presentation. For example, Table 6-1, created by EPA,

---

[1] Probability is a form of uncertainty information.

compares the expected reduction in nonfatal heart attacks in several age groups from two different strategies for attaining national ambient air quality standards, including the 95 percent confidence interval for all values.

Percentage and frequency formats have been found to be more effective than other formats (such as stating that there is a 1-in-X chance of an occurrence) for some circumstances because they more readily allow readers to conduct mathematical operations, such as comparisons, on risk probabilities (Cuite et al., 2008). Other research, however, found that probabilistic reasoning improves and that the influence of cognitive biases (see further discussion below) decreases when information is presented in the form of natural frequencies (for example, 30/1,000) rather than as proportions and single-event probabilities (for example, 3 percent) (see Brase et al., 1998; Gigerenzer, 2002; Gigerenzer and Edwards, 2003; Hoffrage et al., 2000; Kramer and Gigerenzer, 2005). Hoffrage et al. (2000) tested physicians' ability to calculate the predictive value of a screening test for colorectal cancer when information was presented in terms of probabilities—a task that required combining multiple probabilities. Only 1 out of 24 physician participants correctly calculated the false-positive rate when provided data as percentages. In contrast, when provided as fractions (for example, 30/10,000 people), 16 out of 24 of the physicians correctly calculated the false-positive rate (Hoffrage et al., 2000).

Among the disadvantages of numeric presentations are that they are only useful if the people the agency is communicating with are capable of interpreting the numeric information presented and that they may not hold people's attention as well as verbal and graphic presentations (Krupnick et al., 2006; Lipkus, 2007). The appropriateness of such presentations will depend on with whom EPA is communicating. For example, numeric presentations might be more appropriate for EPA decision makers and stakeholders with technical backgrounds than for stakeholders with less technical backgrounds.

As discussed by Peters (2008), decision making is part deliberative (that is, analytical or reason-based) and part affective (that is, intuitive or based on emotional feelings), and using a combination of both approaches is important to good decision making (Damasio, 1994; Slovic and Peters, 2006). The numerical ability, or numeracy, of people varies, however, and this numeracy plays a role in the interpretation of numerical data and in judgments and decisions (Peters et al., 2006). People with higher numeracy are more likely to "retrieve and use appropriate numerical principles and transform numbers presented in one frame to a different frame" (p. 412), and they "tend to draw more affective meaning from probabilities and numerical comparisons than the less numerate [people] do" (pp. 412–413). Both laypeople's and scientist's judgments about risks are often influenced by affective feelings, but the format in which risk data are presented can

BOX 6-1
Strengths and Weaknesses of Numeric, Verbal, and Visual Communication of Risk[a]

Numeric communication of risk (e.g., percentages, frequencies)

Strengths

- Is precise and leads to more accurate perceptions of risk than the use of probability phrases and graphical displays
- Conveys aura of scientific credibility
- Can be converted from one metric to another (e.g., 10% = 1 out of 10)
- Can be verified for accuracy (assuming enough observations)
- Can be computed using algorithms, often based on epidemiological and/or clinical data, to provide a summary score

Weaknesses

- Lacks sensitivity for adequately tapping into and expressing gut-level reactions and intuitions
- People have problems understanding and applying mathematical concepts (level of numeracy)
- Algorithms used to derive numbers may be incorrect, untestable, or result in wide confidence intervals that may affect public trust

Verbal communication of risk (e.g., unlikely, possible, almost, certain)

Strengths

- Allows for fluidity in communication (is easy and natural to use)
- Expresses the level, source, and imprecision of uncertainty, encourages one to think of reasons why an event will or will not occur (i.e., directionality)
- Unlike numbers, may better capture a person's emotions and intuitions

Weaknesses

- Especially if the goal is to achieve precision in risk estimates, variability in interpretation may be a problem (e.g., *likely* may be interpreted as a 60% chance by one person and as an 80% chance by another)

Visual (graphic) communication of risk (e.g., pie charts, scatter plots, line graphs)

Strengths

- Ability to summarize a great deal of data and show patterns in the data that would go undetected using other methods
- Useful for priming automatic mathematic operations (e.g., subtraction in comparing the difference in height between two bars of a histogram)
- Is able to attract and hold people's attention because it displays data in concrete, visual terms
- May be especially useful to help with visualization of part-to-whole relationships

Weaknesses

- Data patterns may discourage people from attending to details (e.g., numbers)
- Poorly designed or complex graphs may not be well understood, and some individuals may lack the skills or educational resources to learn how to use and interpret graphs
- Graphics can sometimes be challenging to prepare or require specialized technical programs
- The design of graphics can mislead by calling attention to certain elements and away from others

[a]The strengths and weaknesses will vary depending on the stage of the decision, the purpose of the communication, and the audience.
SOURCE: Adapted from Lipkus, 2007.

TABLE 6-1 Estimated Reduction in Nonfatal Acute Myocardial Infarctions Associated with Illustrative Attainment Strategies for the Revised and More Stringent Alternative PM NAAQS in 2020

	Reduction in Incidence (95% Confidence Interval)	
Age Interval	15/35 Attainment Strategy	14/35 Attainment Strategy
18–24	1	4
	(1–2)	(2–6)
25–34	8	26
	(4–12)	(13–40)
35–44	170	280
	(84–250)	(140–430)
45–54	520	930
	(260–790)	(460–1,400)
55–64	1,300	2,100
	(630–1,900)	(1,100–3,200)
65–74	1,500	2,600
	(770–2,300)	(1,300–3,900)
75–84	980	1,800
	(490–1,500)	(900–2,800)
85+	520	940
	(260–780)	(460–1,400)
Total	5,000	8,700
	(2,500–7,500)	(4,300–13,000)

NOTE: PM NAAQS = Particulate Matter National Ambient Air Quality Standards.
SOURCE: Modified from EPA, 2008.

affect the interpretation of results to a greater extent in people with low numeracy. For example, low-numeracy individuals perceive risk to be higher when given the information about risk in frequency formats than when given the information in percentage formats (Peters et al., 2011). Presenting information in a manner that facilitates understanding is important, therefore, to people understanding the risks and making decisions using both deliberative and affective approaches (Peters, 2008).

Verbal Presentations of Uncertainty

Verbal presentations of risk—for example, messages containing words such as "likely" or "unlikely"—can be used as calibrations of numeric risk. Such representations may do a better job of capturing people's attention than numeric presentations, and they are also effective for portraying directionality. People are typically familiar with verbal expressions of risk from everyday language (for example, the phrase "It will likely rain tomorrow"), and for some people such presentations may be more user friendly than quantitative portrayals. Furthermore, as discussed by Kloprogge et al.

(2007), verbal expressions of uncertainty can be better adapted to the level of understanding of an individual or group than can numeric and graphic presentations.

A major weakness of verbal or linguistic presentations of risk is that studies have shown that the probabilities attributed to words such as "likely" or "very likely" varies among individuals and can even vary for a single individual depending on the scenario being presented (see Wallsten and Budescu, 1995; Wallsten et al., 1986). For example, as discussed by Morgan (2003), in a study that asked members of the executive committee of EPA's Science Advisory Board about the probabilities attached to the words "likely" and "unlikely" in the context of carcinogenicity, "the minimum probability associated with the word "likely" spans four orders of magnitude, the maximum probability associated with "not likely" spans more than five orders of magnitude," and there was an overlap between the ranges of probabilities associated with the two words (Morgan, 1998, p. 48). That variation can raise a variety of issues when consistency in the interpretation of a health risk is one of the goals of a communication. However, Erev and Cohen (1990) suggested that such vague verbal presentations of information might lead to a consideration of a wider variety of actions within a group, which could be beneficial to the overall group.

Qualitative descriptions of probability—that is, those that include a description or definition for a category of certainty—are sometimes used instead of such subjective calibrations as "very likely" or "unlikely," which are open for individual interpretation. The third assessment report of the Intergovernmental Panel on Climate Change (IPCC), published in 2001 (IPCC, 2001), made extensive use of a qualitative table proposed by Moss and Schneider (2000) as well as of more quantitative likelihood scales (see Table 6-2); these presentations were also used in slightly modified forms in the fourth assessment report published in 2007 (IPCC, 2007). The International Agency for Research on Cancer (IARC) also uses defined categories to classify evidence. For example, IARC classifies the relevant evidence of carcinogenicity from human studies for a given chemical as *limited evidence of carcinogenicity* when "[a] positive association has been observed between exposure to the agent and cancer for which a causal interpretation is considered by the Working Group to be credible, but chance, bias or confounding could not be ruled out with reasonable confidence" (IARC, 2006, p. 19). Such presentations, which provide a description of the state of the science in a given field, can help policy makers with decisions when definitive findings are still pending.

Such use of qualitative likelihood presentations, however, has not been problem-free. Recent research suggests that people may interpret the IPCC qualitative presentations with less precision than intended (Budescu et al., 2009) and that estimates for negatively worded probabilities, such as "very

TABLE 6-2 Supplemental Qualitative Table Used by the Intergovernmental Panel on Climate Change to Describe Its Confidence in Conclusions and Results

Level of Agreement/Consensus			
HIGH		Established but Incomplete	Well Established
		Speculative	Competing Explanations
LOW		Amount of Evidence (Observations, model output, theory, etc.)	HIGH

NOTE: Key to qualitative "state of knowledge" descriptors:
Well Established: Models incorporate known processes, observations largely consistent with models for important variables, or multiple lines of evidence support the finding.
Established but incomplete: Models incorporate most known processes, although some parameterizations may not be well tested; observations are somewhat consistent with theoretical or model results but incomplete; current empirical estimates are well founded, but the possibility of changes in governing processes over time is considerable; or only one or a few lines of evidence support the finding.
Competing Explanations: Different model representations account for different aspects of observations or evidence, or incorporate different aspects of key processes, leading to competing explanations.
Speculative: Conceptually plausible ideas that have not received much attention in the literature or that are laced with difficult to reduce uncertainties or have few available observational tests.
SOURCE: Moss and Schneider, 2000.

unlikely," may be interpreted with greater variability than probability estimates that are positively worded (Smithson et al., 2011). The use of double negatives was especially confounding (Smithson et al., 2011). Budescu et al. (2009) also found that there is interindividual variability in the interpretation of the IPCC categories for certainty, and they recommended using "both verbal terms and numerical values to communicate uncertainty" and adjusting "the width of the numerical ranges to match the uncertainty of the target events" (p. 306). They also recommended describing events as precisely as possible—for example, avoiding the use of subjective terms such as "large"—and specifying the various sources of uncertainty and outlining their type and magnitude.

Graphical Presentation of Uncertainty

Graphical displays of probabilistic information—such as bar charts, pie charts, and line graphs—can summarize more information than other presentations, can capture and hold people's attention, and can show patterns and whole-to-part relationships (Budescu et al., 1988; Spiegelhalter et al., 2011). Furthermore, uncertainties about the outcomes of an analysis can also be depicted using graphical displays, such as bar charts, pie charts, probability density functions,[2] cumulative density functions,[3] and box-and-whisker plots. There is some evidence that graphic displays of uncertainty can help convey uncertainty to people with low numeracy (Peters et al., 2007). A few studies have explored how well different graphical displays of quantitative uncertainty can convey information and have analyzed the effects of different graphical displays on decision making (Bostrom et al., 2008; Visschers and Siegrist, 2008).

Ibrekk and Morgan (1987) compared nine graphical displays by seeing how well each of them communicated univariate uncertainty to 49 well-educated semitechnical and nontechnical people (see Figure 6-1). Participants were asked to estimate the mean, the probability that a value that occurs will be greater than some stated value (that is, x > a), and the probability that the value that occurs will fall within a stated interval (that is, b > x > a). They were first asked to make those estimates without an explanation of how to use or interpret the displays, and then they were asked again after receiving detailed nontechnical explanations. Participants were most accurate with their estimates when they had been shown graphics that explicitly marked the location of the mean (displays 1 and 8), contained the answers to questions about x > a and b > x > a (displays 2 and 9), or provided the 95 percent confidence interval (display 1). In making judgments about best estimates using displays of probability density, subjects tended to select the mode rather than the mean unless the mean was marked. Subjects reported being most familiar with the bar chart and pie chart (displays 2 and 3, respectively), but there was no relationship between familiarity with a display and how sure subjects were of their responses. The researchers also found that participants with some working knowledge of probability and statistics did not perform significantly better in interpreting the displays than participants without such knowledge. One implication of this research is that it will be important to include nontechnical people and people with knowledge of probability, such as EPA decision makers, in research on the communication of uncertainty.

---

[2] Probability density functions show the probability of a given value, for example, the probability that there will be 12 inches of snow.

[3] Cumulative density functions show the probability of something being less than or equal to a given value, for example, the probability that there will be 12 inches of snow or less.

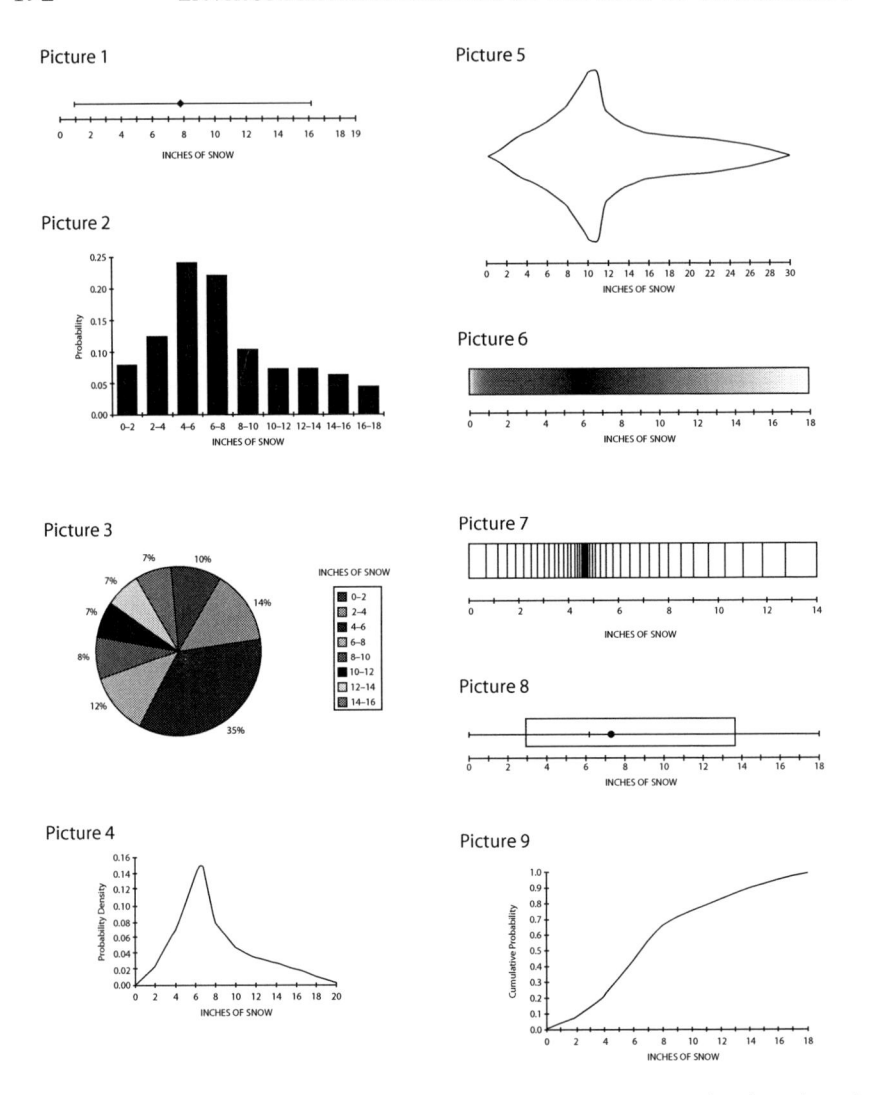

192 *ENVIRONMENTAL DECISIONS IN THE FACE OF UNCERTAINTY*

FIGURE 6-1 Nine displays for communicating uncertain estimates for the value of a single variable used in experiments.
Picture 1: point estimate with an error bar; Picture 2: bar chart; Picture 3: pie chart; Picture 4: conventional probability density function; Picture 5: probability density function of half its regular height together with its mirror image; Picture 6: horizontal bar shaded to display probability density using dots; Picture 7: horizontal bar shaded to display probability density using lines; Picture 8: Tukey box plot modified to exclude the maximum and minimum values and to display the mean with a solid point; Picture 9: conventional cumulative distribution function.
SOURCE: Ibrekk and Morgan, 1987, p. 521. Reprinted with permission of John Wiley & Sons Ltd.

In a small exploratory study, Krupnick et al. (2006) tested the effectiveness of seven different presentations (two tables and five figures)[4] of information about uncertainty in helping seven former high-level EPA decision makers make a decision about whether to adopt a hypothetical proposed tightening of the CAIR.[5] Based on the presentations, decision makers were asked to decide whether they would choose (a) an option to do nothing, (b) an intermediate option of reducing the nitrogen oxide (NO_x) cap by an additional 20 percent below baseline in 2020, or (c) a more stringent option of reducing the NO_x cap 40 percent from baseline in 2020. While acknowledging the small sample size, the authors noted that there were a few findings that deserved to be explored further in future research. One finding was that tables and probability density functions appeared to be best suited for informing high-level decision makers about uncertainty in a scenario of choosing between two regulatory options. Participants found tables to be informative and easy to interpret, and they found probability density functions to be the most familiar of the graphic displays. When asked whether probability density functions might create a bias toward tighter spread, almost all participants reported that the probability density function made them inclined to choose the intermediate option. Participants had more difficulty interpreting the cumulative density function, and most of the participants said that the cumulative density function did not help with decision making.[6] Participants reported that a graphic that listed the sources of uncertainty and described the impact of each source of uncertainty on the estimates of net benefits (see Figure 6-2) was an important input to decision making, gave them insight into how confident they should be in their decision, and prepared them to argue their choice with critics. Participants were able to interpret a pie chart and box-and-whisker plot, but some participants reported that those graphics did not add useful information to what had been provided in the summary tables. The authors were not able to make any generalizations about the impact of the different graphic presentations on decision making; participants' final policy choices were not unanimous, and each decision option—to do nothing,

---

[4] The seven presentations, in order, were (1) a table showing the impacts of the two proposed policies in terms of physical health impacts and costs in 2025; (2) a table with results from a cost–benefit analysis showing total benefits, costs, and net benefits in 2025; (3) a pie chart displaying the probabilities that the policies would produce positive net benefits; (4) a box-and-whisker plot; (5) a probability density function; (6) a cumulative density function; and (7) a graph showing the relative contributions of key variables to the uncertainty associated with the estimate of net benefits.

[5] EPA issued the Clean Air Interstate Rule (CAIR) on March 10, 2005. CAIR covers 28 states in the eastern United States with an order to reduce air pollution by capping emissions of sulfur dioxide (SO_2) and nitrogen oxides (NO_x) (EPA, 2008).

[6] In an earlier small exploratory study, Thompson and Bloom (2000) also found that EPA risk manager participants preferred the PDF format over other graphical displays.

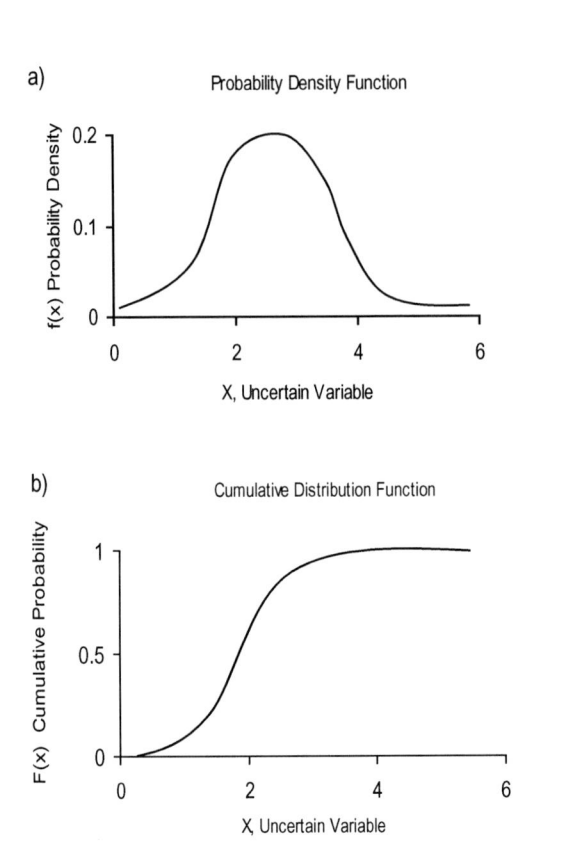

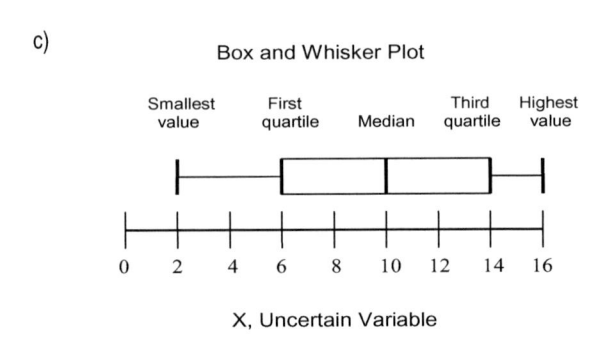

FIGURE 6-2 Examples of the most common graphical displays of uncertainty: (a) a probability density function, (b) a cumulative distribution function, and (c) a box-and-whisker plot.
SOURCE: Adapted from Morgan and Henrion, 1990, p. 221. Reprinted with permission of Cambridge University Press.

COMMUNICATION OF UNCERTAINTY 195

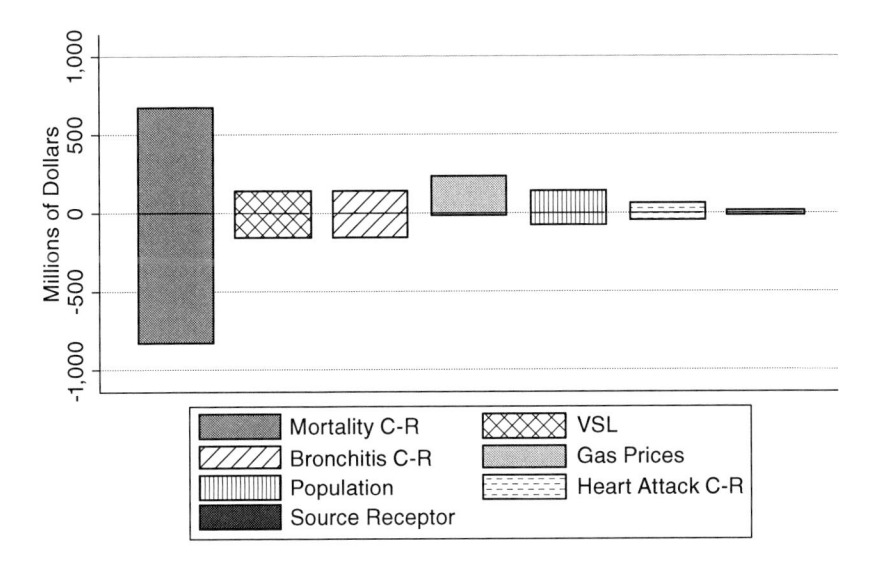

FIGURE 6-3 Graphic used by Krupnick et al. (2006) to display sources of uncertainty and to describe the impact of each source of uncertainty on estimates of expected net benefits in 2025.
NOTES: C–R = concentration–response; VSL = value of statistical lives.
SOURCE: Krupnick et al., 2006. Reprinted with permission of Copyright Clearance Center.

intermediate NO_x cap, and stringent NO_x cap—was selected by at least one study participant.

Krupnick et al. (2006) and Morgan and Henrion (1990) also discussed the strengths and weaknesses of probability and cumulative density functions and the display of selected fractiles, as in box-and-whisker plots (see Figure 6-3). Those are the approaches that are most often used to display uncertainty in probabilistic terms, and each emphasizes different aspects of a probability distribution (Krupnick et al., 2006).[7]

Probability density functions (Figure 6-2a) represent a probability distribution in terms of the area under the curve and highlight the relative probabilities of values. The peak in the curve is the mode, and the shape of

---

[7] Uncertainty along more than one dimension can be graphed using a cumulative distribution function, a probability density function, or a box plot using either multiple graphs or superimposing uncertainties along one dimension over another. An alternative is to use error bars within a line graph, where the error bars represent uncertainty, or to use probability surfaces to depict uncertainty three-dimensionally (Krupnick et al., 2006).

the curve indicates the shape of the distribution (for example, how skewed the data are) (Krupnick et al., 2006). Probability density functions can be a sensitive indicator of variations in probability density, so their use may be advantageous when it is important to emphasize small variations. On the other hand, this sensitivity may sometimes be a disadvantage in that small variations attributed to random sampling may present as noise and are of no intrinsic interest. Another disadvantage may be that the area under the curve, rather than the height of the curve, corresponds to probability. Cumulative distribution functions (CDFs), as illustrated in Figure 6-2b, are calculated by taking the integral of the probability density function, and they best display (1) fractiles (including the median), (2) the probability of intervals, (3) stochastic dominance, and (4) mixed continuous and discrete distributions (Morgan and Henrion, 1990). CDFs have the advantage of not showing as much small variation noise as a probability density function does, so that the shape of the distribution may appear much smoother. As discussed above, however, more people have difficulty interpreting cumulative density functions (Ibrekk and Morgan, 1987). Furthermore, with a cumulative density function it is not as easy to judge the shape of the distribution.

Box-and-whisker plots (Figure 6-2c) are effective in displaying summary statistics (medians, ranges, fractiles), but they provide no information about the shape of the distribution except for the presence of asymmetry in the distribution (Krupnick et al., 2006). The first quartile (the left-hand side of the box) represents the median of the lower part of the data, the second quartile (the line through the middle of the box) is the median of all data, and the third quartile (the right-hand side of the box) is the median of the upper part of the data. The ends of the "whiskers" show the smallest and largest data points.

Ibrekk and Morgan (1987) concluded that, until future research suggests another strategy is more effective, it may be best to use both cumulative and probability density functions with the same horizontal scale and with the location of the mean clearly indicated on each. The decision of which of these displays to use depends on what type of information the user needs to extract, so it is important to understand the information needs of the people the agency is communicating with. One drawback of both density functions, however, is that people—especially people without a strong technical background—may have difficulty extracting summary information.

Despite their advantages, graphic displays do not always explicitly describe conclusions, and they can require more effort to extract information, particularly for people who are not familiar with the mode of presentation or who lack skills in interpreting graphs or in cases where the graphic presents complex data (Kloprogge et al., 2007; Lipkus, 2007).

The interpretation of a graph depends on the "viewer's familiarity with the content depicted in a graph, and the viewer's graphicacy skills" as well as the design of the graph (Shah and Freedman, 2009). For example, graph viewers are less likely to discern the relevant results from a graph if the data in the graph are not grouped to form appropriate "visual chunks" (Shah et al., 1999) or exhibited in a format that supports the intended inferences (Shah and Freedman, 2009). In addition, some research indicates that individuals differ in how they use the information in different presentations of data (Boduroglu and Shah, 2009), further complicating the use of graphical presentations.

There is also some evidence that graphic displays increase risk aversion. For example, one study that examined how well visual displays of risk communicated low-probability events found that adding graphics to numeric presentations increased participants' willingness to pay for risk reductions (Stone et al., 1997). There is no correct level of perceived risk, however, so it is not possible to rank the effectiveness of various displays based on this outcome.

Furthermore, graphs can be designed—either intentionally or unintentionally—to call attention to certain aspects of a message and detract from others. Highlighting the foreground rather than the background can make people more risk averse. For example, people are more risk averse after seeing a bar graph that only shows the differences in the number of people suffering from serious gum disease with the denominator of "per 5,000" people included in the figure legend than after seeing a bar graph that depicts both differences in gum disease and the denominator of 5,000 people (Stone et al., 2003). Even when such foreground–background salience and gain–loss framing (see discussion below) are controlled, however, evidence indicates that graphic displays lead to greater risk aversion than numerical presentations (Slovic and Monahan, 1995; Slovic et al., 2000).

The ability to use interactive visualizations to display information and uncertainty about that information has increased with the evolution of computer technology. Spiegelhalter et al. (2011) point out that "increasing availability of online data and public interest in quantitative information has led to a golden age of infographics" (p. 1399), including the ability to create graphics with interactive features. Such interactive graphics have the potential to increase understanding and retention and to help counteract differences in numeracy, and this potential could be applied in the communication of uncertainty. Spiegelhater notes, however, that while there is huge potential applications and uses for infographics with interactive features, such graphics have not yet been evaluated empirically.

One limitation of most of those graphical presentations is that they display only one variable at a time. For example, they might show how the uncertainty in an estimate of human health risks varies among individuals with different sensitivities to a chemical or show the consequences of different regulatory decisions on human health benefits. In reality, however, most of the problems that EPA faces have many sources of uncertainty and many intermediate outputs which may covary. For example, the consideration of a number of different health endpoints might influence a decision, or there might be estimates for a number of costs and benefits of a rule, each of which has uncertainty associated with it. Tornado diagrams provide "a pictorial representation of the contribution of each input variable to the output of the decision making model" (Daradkeh et al., 2010).

Framing Biases

One line of risk perception research that is relevant to EPA's communication of uncertainty is the study of the effects that alternative ways of framing risk information have on risk perception and decision making. Experts have been found to be just as susceptible to framing effects as isthe general population (Slovic et al., 1982). Different ways of framing probabilistic information can leave people with different impressions about a risk estimate and, consequently, the confidence in that estimate. For example, stating that "10 percent of bladder cancer deaths in the population can be attributed to arsenic in the water supply" may leave a different impression than stating that "90 percent of bladder cancer deaths in the population can be attributed to factors other than arsenic in the water supply," even though both statements contain the same information. Choices based on presentations of a range of uncertainty will be similarly influenced by the way that information is presented. Risk estimates that include a wide range of uncertainties may imply that an adverse outcome is possible, even if the likelihood of the adverse outcome occurring is extremely small (NRC, 1989).

CONSIDERATIONS WHEN DECIDING ON A COMMUNICATIONS APPROACH

Determining the best approach to communicate the uncertainty in a decision needs to be made on a case-by-case basis (Fagerlin et al., 2007; Lipkus, 2007; Nelson et al., 2009; Spiegelhalter et al., 2011; Visschers et al., 2009). There are, however, a number of considerations that should be taken into account when making that decision. The committee discusses the following considerations below: (1) the stage of the decision-making process and the purpose of the communication; (2) the decision context; (3) the type and source of the uncertainty and the whether the uncertainty analysis is qualitative or quantitative; and (4) the audience with which EPA is communicating. Testing and evaluating the effectiveness of communication approaches is also important (Fischhoff et al., 2011).

Both the U.S. National Institutes of Health (NCI, 2011)[8] and the Netherlands Environmental Assessment Agency have developed guidance on communicating uncertainty (Kloprogge et al., 2007). Although both guidance documents emphasize the need for communication strategies to be developed on a case-by-case basis, they also present generalities about the strengths and weaknesses of different approaches and describe some circumstances under which one approach might be preferable over another. NIH's workbook includes some detailed suggestions, such as the order in which to present data and color choices (NCI, 2011).

The Stage of the Decision-Making Process and the Purpose of the Communication

The most appropriate strategy for communicating uncertainty will depend in part on the phase of the decision-making process and the purpose of the communication. Chapter 5 identified three phases in the decision-making process: problem formulation, assessment, and implementation. The key to a good communication strategy is initiating communication during the problem-formulation phase and continuing it throughout the decision-making process. The purpose of the communication, however, might differ from one phase to the next.

During the problem-formulation phase, the communication strategy should ensure input from stakeholders on what uncertainties they are aware of and concerned about and on how those uncertainties should be accounted for in the assessment and implementation phases. A key goal of communication about uncertainty during the problem-formulation phase is to develop a common understanding of the decision problem, of the limits or constraints on the decision options, and of the potential uncertainties that exist in the evidence base for the decision.

The understanding gained from the problem-formulation phase will help shape the assessments that occur during the second phase of the decision-making process. Further communication might be needed to clarify issues about uncertainties and to discuss any new uncertainties that are identified during the assessment and how those uncertainties should be considered in the assessment.

During the implementation phase, the assessors will characterize the risks, costs and benefits, and other factors that were assessed during the

---

[8] The National Institutes of Health's workbook operationalizes the main points contained in the book *Making Data Talk: Communicating Public Health Data to the Public, Policy Makers, and the Press* (Nelson et al., 2009).

assessment phase, with EPA's decision makers being an important audience at this stage. Those characterizations should include a characterization of the uncertainties in the data and analyses that underlie each of the factors that were assessed. One type of communication during the implementation phase will be the agency communicating with stakeholders to discuss its decision and the rationale for its decision, including how uncertainties affect the decision. Another part of the communication process should be the agency getting feedback on the decision and uncertainties as well as having discussions about how and when the decision will be revisited.

Decision Context

To develop guidance on communicating uncertainty, the Netherlands Environmental Assessment Agency relied on literature reviews, a workshop on uncertainty communication, and research by the authors of the guidance document (Kloprogge et al., 2007). The guidance emphasizes the importance of decision context. The context of a decision—that is, the characteristics of the setting in which the decision is being made—affects the communication of the decision and of the uncertainty underlying it. Box 6-2 lists some decision contexts in which the communication of the uncertainty surrounding a decision is particularly important. A complex decision based on controversial science or a decision about which stakeholders disagree will benefit from greater attention to communicating uncertainties (Kloprogge et al., 2007).

It can be especially challenging to communicate the uncertainty associated with a decision made in an emergency situation, such as a hazardous chemical spill. Under such circumstances, EPA must communicate not only with those involved in containment and cleanup, but also with members of the public who might be affected by the spill, and the communication may need to be done in coordination with other agencies, with governments, and with stakeholders such as private companies involved in the spill. Such communication, sometimes called *crisis communication*, is often carried out at a time when there are a number of large uncertainties about the event and its potential consequences on human health and the environment (Reynolds and Matthew, 2005). The time frame within which a decision is needed in an emergency situation can limit the time and opportunities available for communication, and the purpose of communication in such a situation can differ from traditional risk communications in that crisis communication often is principally informative (Reynolds and Matthew, 2005). Although it is generally not possible to predict the timing and extent of an emergency, the nature of many potential emergencies can, and often are, known and planned for. Communicating with stakeholders about the uncertainties that might follow an emergency during the planning for such an emergency and

> **BOX 6-2**
> **When Greater Attention to Reporting Uncertainties May Be Needed**
>
> Reporting uncertainties may be more policy relevant when
>
> - The outcomes are very uncertain and have a great impact on the policy advice given.
> - The outcomes are situated near a policy target, threshold, or standard set by policy.
> - A wrong estimate in one direction will have entirely different consequences for policy advice than a wrong estimate in another direction.
> - There is a possibility of morally unacceptable damage or "catastrophic events."
> - Controversies among stakeholders are involved.
> - There are value-laden choices or assumptions that are in conflict with the views and interests of stakeholders.
>
> Greater attention to reporting uncertainties may also be needed when
>
> - Fright factors or media triggers are involved.
> - There are persistent misunderstandings among audiences.
> - The audiences are expected to distrust outcomes that point to low risks because the public perception is that the risks are serious.
> - The audiences are likely to distrust the results because of low or fragile confidence in the researchers or the organization that performed the assessment.
>
> SOURCE: Kloprogge et al., 2007.

explicitly including the communication of uncertainties in emergency plans can help facilitate communications when an environmental crisis requiring an emergency response occurs. Given the need for a quick decision and the large amount of uncertainty that often occurs in emergency situations, it is important that communication strategies include plans to collect information that might reduce uncertainties or plans to revisit the decision once more data are gathered.

The decision context could also determine whom the agency and its technical staff should communicate with. Furthermore, as discussed below, the characteristics of those with whom EPA is communicating should also affect the strategy for communicating the decision, including the uncertainty in the decision. The communication strategy for a decision that will affect only a small region will differ from the communication strategy for a decision that will have consequences on a national scale. A decision might also have a greater effect on one subgroup than another (for example, a decision that affects the levels of a chemical in fish might affect anglers more than other people), and those subgroups that are more at risk should be identified during the problem-formulation phase, and discussions about potential uncertainties should be initiated during that phase.

The Type and Source of the Uncertainty

Some research indicates that, when communicating uncertainties in the results of risk assessments to decision makers, it is valuable to be specific about the nature or types of the uncertainties. Bier (2001b) discusses two types of outcome uncertainties: (1) state of knowledge or assessment uncertainty, and (2) variability or randomness (in other words, uncertainties arising from variability or natural variation elements in such factors as environments, populations and exposure paths), which cannot be controlled and are thus not reducible.[9] Those two categories of uncertainties correspond to what the committee refers to as model and parameter uncertainty, and variability and heterogeneity, respectively.[10] Bier (2001b) suggests that, when communicating uncertainties to decision makers, it is helpful to distinguish between the two types of the uncertainties so that the decision maker can understand how much of the uncertainty in the decision may be reducible. For example, if it is not possible to wait for research to reduce state-of-knowledge uncertainty, a decision maker may give more priority to a risk for which there is large state-of-knowledge uncertainty and a small population variability rather than to a risk for which there is large population variability and small state-of-knowledge uncertainty (Bier, 2001b). Others have argued, however, that in most instances this distinction is not a useful one to make since it can result in an overly complicated and confusing analysis (Morgan and Henrion, 1990). It is also important to communicate the sources of uncertainty—for example, whether it arises from the estimates of human health, estimates of costs, the availability of technology, or other factors—and to include the relative impact of the different sources on the decision. Such communication should also discuss the results of any sensitivity analyses, so that the uncertainty is bounded.

---

[9] State-of-knowledge uncertainties are uncertainties due to limited scientific knowledge about the models that link causes and effects of risk and risk-reduction actions as well as about the specific parameters of these models. Variability refers to natural variation elements (environments, populations, exposure paths, etc.) that cannot be controlled.

[10] The committee also separates out deep uncertainty, which is not immediately reducible through research (see Chapter 1 for a discussion).

Technical experts often communicate uncertainties to decision makers in aggregate. However, aggregate estimates of uncertainty do not necessarily provide the decision makers with an understanding of the uncertainty and its implications for a decision. Information about the type and sources of uncertainty can help decision makers decide whether further research is warranted to decrease the uncertainty or whether to refine the decision to reduce the effects of the uncertainty. Descriptions of where there are uncertainties can also indicate which groups or stakeholders might bear the burden of a higher-than-anticipated health risk or cost because of the uncertainty. Knowing who is likely to be affected by the uncertainty in the costs and benefits would allow decision makers to design the initial proposed regulations to address or prepare for those potential outcomes in advance. If the individual sources of uncertainty that contribute to the overall uncertainty can be determined, then uncertainty analyses in, for instance, cost–benefit analyses or cost-effectiveness analyses could incorporate graphic representations displaying the relative importance of the different sources of uncertainty sequentially so as to provide an easily interpretable graphic display of the sources of uncertainty (Krupnick et al., 2006).

Audience Characteristics

Level of Technical Expertise

The audience for the communication of uncertainty in environmental decisions, such as those made by the EPA, will have a broad array of backgrounds and roles (Wardekker et al., 2008). EPA's scientific and technical staff communicates about uncertainty in health risk estimates, economic analyses, and other factors with agency decision makers. The agency discusses the uncertainties in its decisions with stakeholders, including individuals who might have little to no technical knowledge, as well as with industry specialists and others with high levels of technical expertise. The uncertainty that affects decisions should be discussed with all stakeholders, but the strategy used for those discussions might vary with the technical expertise of the audience. For example, agency decision makers will often have strong technical backgrounds and might need to see specific numbers to best understand the extent of uncertainty and how it affects their decisions. Industry and advocacy group scientists similarly might prefer specific numbers, as such numbers might provide them with a complete picture and the data needed for them to conduct their own independent analyses. Members of the public without strong technical backgrounds might benefit more from graphic representations of the uncertainties along with discussions about how those uncertainties will be considered in a decision, the potential consequences of those uncertainties, and whether and how EPA plans to decrease those uncertainties. Regardless of the audience, however, EPA should use its communication opportunities to provide audiences with information as well as to gather information from the audience that could help either decrease acknowledged uncertainties or identify additional uncertainties that might affect the decision.

Another potential option now that many documents are available electronically through the Internet is to use layered hypertext for more complex uncertainty analysis. That is, the main body of text and the summary sections of EPA's decision document could contain a summary of the uncertainty analyses conducted and could also include a link to appendixes or other documents that present full details of the analyses. That would provide summary information for all audiences as well as further details of the uncertainty analyses for technical audiences or others with an interest in seeing all the details.

Biases

Uncertainty information concerning probabilities has been found to be susceptible to biases by both experts and non-experts (Hoffrage et al., 2000; Kloprogge et al., 2007; Slovic, 2000; Slovic et al., 1979, 1981; Tversky and Kahneman, 1974). When people's judgments about a risk are biased, risk management and communication efforts may not be as effective as they would otherwise be. Biases can stem from the characteristics of an individual or group or can be embedded in the framing of a message; both types can influence the interpretation of a message. Communicators of information about uncertainty cannot completely eliminate these biases, but they should be aware of the potential for biases to influence the acceptance of and reaction to probabilistic information and, to the extent possible, account for these biases by adjusting communication efforts. These types of biases are discussed below.

Personal Biases One bias that can affect how people interpret probabilistic information is termed availability bias. People tend to judge events that are easily recalled as more risky or more likely to occur than events that are not readily available to memory (see Kloprogge et al., 2007; Slovic et al., 1979; Tversky and Kahneman, 1974). An event may have more availability if it occurred recently, if it was a high-profile event, or if it has some other significance for an individual or group. The overestimation of rare causes of death that have been sensationalized by the media is an example of availability bias. One implication of availability bias that communicators of risk and uncertainty information should be aware of is that the discussion of a risk may increase its perceived riskiness, regardless of what the actual risk may be (Kloprogge et al., 2007). For example, evidence indicates that women overestimate their risk of having breast cancer; women believe that their risk of breast cancer is higher than their risk of cardiovascular disease, despite the fact that cardiovascular disease affects and kills more women than breast cancer (Blanchard et al., 2002).

A second bias that can influence the communication of health risks and their uncertainties is confirmation bias. Confirmation bias refers to the filtering of new information to fit previously formed views; in particular, it is the tendency to accept as reliable new information that supports existing views, but to see as unreliable or erroneous and filter out new information that is contrary to current views (Russo et al., 1996). People may ignore or dismiss uncertainty information if it contradicts their current beliefs (Kloprogge et al., 2007). Evidence indicates that probability judgments are subject to confirmation bias (Smithson et al., 2011). Communicators of risk information, therefore, should be aware that peoples' preexisting views about a risk, particularly when those views are very strong, may be difficult to change even with what some would consider to be "convincing" evidence with little uncertainty.

A third bias is confidence bias. People have a tendency to be overconfident about the judgments they make based on the use of heuristics. When people judge how well they know an uncertain quantity, they may set the range of their uncertainty too narrowly (Morgan, 2009). Research by Moore and Cain (2007) supports the notion that people may overestimate or underestimate their judgments based on their level of confidence. Referred to as the overconfidence bias, this tendency seems to have its basis in a psychological insensitivity to questioning of the assumptions upon which judgments are based (Slovic et al., 1979, 1981).

Group Biases The literature on public participation emphasizes the importance of interaction among stakeholders as a way of minimizing the cognitive biases that shape how people react to risk information (see Renn, 1999, 2004). Kerr and Tindale (2004), for example, caution that the more homogeneous a group is with respect to knowledge and preferences, the more strongly the knowledge and preferences will affect a group decision. Uncertainty can be either amplified or downplayed, depending on a group's biases toward the evidence.

Assessment and explicit acknowledgement of the biases of the people that the agency is communicating with might be critical to successful communication. People may be more willing to listen to new information and other points of view after their own concerns have been acknowledged and validated (Bier, 2001a).

EPA's scientists and technical staff are themselves not immune to these biases. An awareness of the possible biases within EPA and when they occur would be a first step toward identifying biases and helping decrease the possibility that such biases influence the interpretation and presentation of scientific evidence.

Considerations for Communicating with Journalists

The statement of task asks the committee if there are specific communication techniques that could improve understanding of uncertainty among journalists. This is an important question, as most members of the public get their information about risks from the media. Journalists and the media help to identify conflicts about risk, and they can be channels of information during the resolution of those conflicts (NRC, 1989). Journalists do generally care about news accuracy and objectivity (NRC, 1989; Sandman, 1986) and about balance in representation of opinions, but journalists vary widely in their backgrounds, technical expertise, and ability to accurately report and explain environmental decisions. Even those who cover environmental policy making will not necessarily be familiar with the details of risk assessment and its inherent uncertainties, making it challenging to convey the rationale for decisions based, in part, on those assessments.

Uncertainty is not unique to reporting on environmental health risks, of course. Studies of how the U.S. news media handle uncertainty in science in general have found that journalists tend to make science appear more certain and solid than it is (see Fahnestock, 1986; Singer and Endreny, 1993; Weiss et al., 1988). In a quantitative content analysis, for example, Singer and Endreny (1993) found that the media tended to minimize uncertainties of the risks associated with natural and manmade hazards. The issue of which factors might contribute to this tendency to minimize uncertainties has not yet been studied, but the tendency could be related to journalists' understanding of uncertain information versus their incentive to develop attention-grabbing stories that omit or downplay uncertainties. It should be expected that journalists, just like most other people, will tend to interpret risk messages based on their existing beliefs. The reporting of risk and uncertainty information in the media will be influenced accordingly.

Because the journalists and the media are a major avenue for framing risk information and its inherent uncertainty, efforts are needed to ensure that they are well informed of what is known about risks and risk-management options, including the sources and magnitude of uncertainty and its implications; a particularly useful approach would be to provide journalists with short, concise summaries about those implications. Although such summaries can be a challenge to develop, it can be done. For example, as discussed in Chapter 2, the summary of the regulatory impact analysis for the CAIR (EPA, 2005) contains a summary discussion of the uncertainty analysis. Those who are most familiar with the risk and uncertainties should provide the perspective that the journalists seek and should recognize the limitations and constraints of the media. Although little research has been carried out on the best means of providing journalists with such a perspective, providing agency personnel with training on how to communicate effectively with media representatives about uncertainties may prove helpful to journalists, as might providing journalists with access to the agency officials who were involved in the decision making. Providing the media with summaries of the uncertainties in the risk assessment and risk management in a variety of formats may also help ensure that the uncertainties are conveyed accurately.

Social Trust

An important concept related to stakeholder values and perceptions is social trust. Trust has long been considered of central importance to risk management and communication (Earle, 2010; Earle et al., 2007; Kasperson et al., 1992; Löfstedt, 2009; Renn and Levine, 1991). Slovic (1993) noted an inverse relationship between the level of trust in decision makers and the public's concern about or perception of a risk—that is, the lower the trust, the higher the perception of risk. The importance of organizational reputation is not unique to EPA; in *Reputation and Power: Organizational Image and Pharmaceutical Regulation at the FDA*, Carpenter (2010) emphasized the importance that the U.S. Food and Drug Administration's reputation plays in its regulatory authority.

Frewer and Salter (2012) point out that beliefs about the underlying causes of trust or distrust and about the best approaches for increasing trust have changed over the past few decades. In contrast to the old idea that increasing knowledge will increase trust, Frewer et al. (1996) found that certain inherent aspects of the source of information—such as having a good track record, being truthful, having a history of being concerned with public welfare, and being seen as knowledgeable—lead to increased trust. Similarly, Peters et al. (1997) found that the source of the information being seen as having "knowledge and expertise, honesty and openness, and concern and care" was an important contributor to trust (p. 10). In a study looking at attitudes toward genetically modified foods, however, Frewer et al. (2003) found that neither the information itself nor the strategy for communicating the risks had much effect on people's attitudes toward genetically modified foods; in this case, people's attitudes toward genetically modified foods tended to determine their level of trust in the source of information, rather than the trust in the source determining their attitudes toward the foods. It is important to remember, however, that there are reasons to communicate uncertainties beyond the potential to increase social trust (Stirling, 2010).

Earle (2010) reviewed the distinction between trust, which is about relationships between people, and confidence, which concerns a relationship between people and objects, and the role of both in social trust. As Earle (2010) pointed out, although some people believe that decisions should be made on the basis of data or numbers (Baron, 1998; Bazerman et al., 2002; Sunstein, 2005), any "confidence-based approach presupposes a relation of trust" (p. 570). For decisions concerning hazards of high moral importance, that trust does not necessarily exist (Earle, 2010).

As discussed by Fischhoff (1995) and by Leiss (1996), at earlier stages in the evolution of risk communication sciences it was thought that public education via increased communication would lead to an increased understanding of the concept of risk and, subsequently, to increased trust. Furthermore, some research had indicated that a decrease in public confidence in regulatory agencies and scientific institutions—and their motives—led to decreased trust (Frewer and Salter, 2002; Pew Research Center for the People and the Press, 2010). Given those observations, it was thought that increasing transparency would be one way to increase trust. As discussed by Frewer and Salter (2012), however, there is limited evidence that transparency actually does increase trust, although there is evidence that a lack of transparency can lead to increased distrust (Frewer et al., 1996). As highlighted in the discussion of the committee's framework for decision making in Chapter 5, all aspects of the decision-making process, including the more technical risk assessment process, require value judgments. Thus engaging the public and policy makers, in addition to scientists, in the process of health risk assessments not only improves the assessment, but can also increase both trust in the process and communications about health risks by allowing the perspectives of all stakeholders to inform the assessment. Frewer and Salter (2002) described the communications by the United Kingdom's regulatory agencies related to the bovine spongiform encephalopathy (BSE) outbreak in the mid-1990s in the United Kingdom as an example of the consequences of inadequate public participation in the decision process. Communications about the outbreak and the outbreak response did not address many of the concerns of the public and led to public outrage about the response.

Concerning the communication of uncertainties in risks, Frewer and Salter (2012) pointed out that distrust in risks assessments will increase when uncertainties are not included in the discussion of the assessments. Although some researchers noted, for the BSE outbreak in the United Kingdom, an apparent view by government officials "that the public [is] unable to conceptualize uncertainty" (Frewer et al., 2002, p. 363), research on risks related to food safety indicates a preference by the public to be informed of uncertainties in risk information (Frewer et al., 2002) and finds that not discussing uncertainties "increases public distrust in institutional activities designed to manage risk" (Frewer and Salter, 2012, p. 153).

Although, there is insufficient information to develop guidelines or best practices for communicating the uncertainty and variability in health risk estimates (Frewer and Salter, 2012), there is evidence that the public can differentiate between different types and sources of uncertainty (see below for further discussion). As discussed by Kloprogge et al. (2007), it is possible to communicate to the public various aspects of uncertainty information, such as how uncertainty was dealt with in the analysis as well as the implications of uncertainties and what can or cannot be done about uncertainties.

The need for a communication plan is increased when there are—or are expected to be—more uncertainties associated with a decision-making process, because there are likely to be more challenges in communicating with stakeholders. Research demonstrates a heightened interest by the public in evaluating the credibility of information sources when they perceive uncertainty (Brashers, 2001; Halfacre et al., 2000; van den Bos, 2001), and studies also indicate that the public is more likely to challenge the reliability and adequacy of risk estimates and be less accepting of reassurances in the presence of uncertainty (Kroll-Smith and Couch, 1991; Rich et al., 1995). Concerns about procedural fairness and trust appear to be even more salient when there is scientific uncertainty (NRC, 2008), and risk communication can serve to facilitate stakeholder trust (Conchie and Burns, 2008; Heath et al., 1998; Peters et al., 1997).

KEY FINDINGS

- Although communication is often thought of in terms of communication to an audience, two-way conversations about risks and uncertainties throughout the decision-making process are key to the informed environmental decisions that are acceptable to stakeholders. Not only will such communication inform the public and others about decisions, but it will also help to ensure that the decisions take the concerns of various stakeholders into consideration, and to build social trust and broader acceptance of decisions.

RECOMMENDATION 8.1

U.S. Environmental Protection Agency senior managers should be transparent in communicating the basis of the agency's decisions, including the extent to which uncertainty may have influenced decisions.

- There is no definitive research that can serve as a basis for uniform recommendations as to the best approaches to communicating uncertainty information with all stakeholders. Each situation will likely require a unique communication strategy, determined on a case-by-case basis, and in each case it may require research to determine the most appropriate approach. Communicating decision made in the presence of deep uncertainty is particularly challenging.

RECOMMENDATION 8.2
U.S. Environmental Protection Agency decision documents and communications to the public should include a discussion of which uncertainties are and are not reducible in the near term. The implications of each to policy making should be provided in other communication documents when it might be useful for readers.

- The best presentation style will depend on the audience and their needs. When communicating with decision makers, for example, because of the problem of variability in interpretation of verbal presentations, such presentations should be accompanied by a numeric representation. When communicating with individuals with limited numeracy or with a variety of stakeholders, providing numeric presentations of uncertainty may be insufficient. Often a combination of numeric, verbal, and graphic displays of uncertainty information may be the best option. In general, however, the most appropriate communication strategy for uncertainty depends on
  o the decision context;
  o the purpose of the communication;
  o the type of uncertainty; and
  o the characteristics of the audience, including the level of technical expertise, personal and group biases, and the level of social trust.

RECOMMENDATION 9.1
The U.S. Environmental Protection Agency, alone or in collaboration with other relevant agencies, should fund or conduct research on communication of uncertainties for different types of decisions and to different audiences, develop a compilation of best practices, and systematically evaluate its communications.

- Little research has been conducted on communicating the uncertainty associated with technological or economic factors that play a role in environmental decisions, or other influences on decisions that are less readily quantified, such as social factors (for example, environmental justice) and the political context.

RECOMMENDATION 9.2
As part of an initiative evaluating uncertainties in public sentiment and communication, U.S. Environmental Protection Agency senior managers should assess agency expertise in the social and behavioral sciences (for example, communication, decision analysis, and economics), and ensure it is adequate to implement the recommendations in this report.

REFERENCES

Baron, J. 1998. *Judgment misguided: Intuition and error in public decision making.* New York: Oxford University Press.
Bazelon, D. L. 1974. The perils of wizardry. *American Journal of Psychiatry* 131(12):1317–1322.
Bazerman, M. H., J. Baron, and K. Shonk. 2002. *"You can't enlarge the pie": The psychology of ineffective government.* Cambridge, MA: Basic Books.
Bier, V. M. 2001a. On the state of the art: Risk communication to decision-makers. *Reliability Engineering and System Safety* 71:151–157.
———. 2001b. On the state of the art: Risk communication to the public. *Reliability Engineering and System Safety* 71:139–150.
Blanchard, D., J. Erblich, G. H. Montgomery, and D. H. Bovbjerg. 2002. Read all about it: The over-representation of breast cancer in popular magazines. *Preventive Medicine* 35(4):343–348.
Boduroglu, A., and P. Shah. 2009. Effects of spatial configurations on visual change detection: An account of bias changes. *Memory and Cognition* 37(8):1120–1131.
Bostrom, A., and R. E. Löfstedt. 2003. Communicating risk: Wireless and hardwired. *Risk Analysis* 23(2):241–248.
Bostrom, A., L. Anselin, and J. Farris. 2008. Visualizing seismic risk and uncertainty. *Annals of the New York Academy of Sciences* 1128(1):29–40.
Brase, G. L., L. Cosmides, and J. Tooby. 1998. Individuation, counting, and statistical inference: The role of frequency and whole-object representations in judgment under uncertainty. *Journal of Experimental Psychology: General* 127(1):3–21.
Brashers, D. E. 2001. Communication and uncertainty management. *Journal of Communication* 51(3):477–497.
Budescu, D. V., S. Weinberg, and T. S. Wallsten. 1988. Decisions based on numerically and verbally expressed uncertainties. *Journal of Experimental Psychology: Human Perception and Performance* 14(2):281–294.
Budescu, D. V., S. Broomell, and H. H. Por. 2009. Improving communication of uncertainty in the reports of the Intergovernmental Panel on Climate Change. *Psychological Science* 20(3):299–308.
Carpenter, D. P. 2010. *Reputation and power: Organizational image and pharmaceutical regulation at the FDA.* Princeton, NJ: Princeton University Press.
Conchie, S. M., and C. Burns. 2008. Trust and risk communication in high-risk organizations: A test of principles from social risk research. *Risk Analysis* 28(1):141–149.
Covello, V. T., and F. W. Allen. 1988. *Seven cardinal rules of risk communication.* Washington, DC: EPA.
Cuite, C. L., N. D. Weinstein, K. Emmons, and G. Colditz. 2008. A test of numeric formats for communicating risk probabilities. *Medical Decision Making* 28(3):377–384.
Damasio, A. R. 1994. *Descartes' error: Emotion, reason, and the human brain.* New York: Putnam.

Daradkeh, M., A. E. McKinnon, and C. Churcher. 2010. Visualisation tools for exploring the uncertainty–risk relationship in the decision-making process: A preliminary empirical evaluation. *User Interfaces* 106:42–51.
Earle, T. C. 2010. Trust in risk management: A model-based review of empirical research. *Risk Analysis* 30(4):541–574.
Earle, T. C., M. Siegrist, and H. Gutscher. 2007. Trust, risk perception and the TCC model of cooperation. In *Trust in cooperative risk management: Uncertainty and scepticism in the public mind*, edited by M. Siegrist, T. C. Earle, and H. Gutscher. London: Earthscan. Pp. 1–49.
EPA (U.S. Environmental Protection Agency). 2004. *An examination of EPA risk assessment principles and practices*. Washington, DC: Office of the Science Advisor, Environmental Protection Agency.
———. 2005. *Regulatory impact analysis for the final Clean Air Interstate Rule*. Washington, DC: EPA, Office of Air and Radiation.
———. 2007. *Risk communication in action: The risk communication workbook*. Washington, DC: EPA.
———. 2008. Clean Air Interstate Rule. http://www.epa.gov/cair (accessed September 10, 2008).
Erev, I., and B. L. Cohen. 1990. Verbal versus numerical probabilities: Efficiency, biases, and the preference paradox. *Organizational Behavior and Human Decision Processes* 45(1):1–18.
Fagerlin, A., P. A. Ubel, D. M. Smith, and B. J. Zikmund–Fisher. 2007. Making numbers matter: Present and future research in risk communication. *American Journal of Health Behavior* 31(Suppl 1):S47–S56.
Fahnestock, J. 1986. Accommodating science: The rhetorical life of scientific facts. *Written Communication* 3(3):275–296.
Fischhoff, B. 1995. Risk perception and communication unplugged: Twenty years of process. *Risk Analysis* 15(2):137–145.
Fischhoff, B., N. T. Brewer, and J. S. Downs, eds. 2011. *Communicating risks and benefits: An evidence-based user's guide*: Washington, DC: Food and Drug Administration, U.S. Department of Health and Human Services.
Frewer, L., and B. Salter. 2002. Public attitudes, scientific advice and the politics of regulatory policy: The case of BSE. *Science and Public Policy* 29(2):137–145.
———. 2012. Societal trust in risk analysis: Implications for the interface of risk assessment and risk management. In *Trust in risk management: Uncertainty and skepticism in the public mind*, edited by M. Siegrist, T. C. Earle, and H. Gutscher. London: Earthscan. Pp. 143-158.
Frewer, L. J., C. Howard, D. Hedderley, and R. Shepherd. 1996. What determines trust in information about food-related risks? Underlying psychological constructs. *Risk Analysis* 16(4):473–486.
Frewer, L., S. Miles, M. Brennan, S. Kuznesof, M. Ness, and C. Ritson. 2002. Public preferences for informed choice under conditions of risk uncertainty. *Public Understanding of Science* 11:363–372.
Frewer, L. J., J. Scholderer, and L. Bredahl. 2003. Communicating about the risks and benefits of genetically modified foods: The mediating role of trust. *Risk Analysis* 23(6):1117–1133.
Gigerenzer, G. 2002. *Calculated risks: How to know when numbers deceive you*. New York: Simon and Schuster.
Gigerenzer, G., and A. Edwards. 2003. Simple tools for understanding risks: From innumeracy to insight. *British Medical Journal* 327(7417):741–744.

Halfacre, A. C., A. R. Matheny, and W. A. Rosenbaum. 2000. Regulating contested local hazards: Is constructive dialogue possible among participants in community risk management? *Policy Studies Journal* 28(3):648-667.

Heath, R. L., S. Seshadri, and J. Lee. 1998. Risk communication: A two-community analysis of proximity, dread, trust, involvement, uncertainty, openness/accessibility, and knowledge on support/opposition toward chemical companies. *Journal of Public Relations Research* 10(1):35-56.

Hoffrage, U., S. Lindsey, R. Hertwig, and G. Gigerenzer. 2000. Medicine: Communicating statistical information. *Science* 290(5500):2261-2262.

IARC (International Agency for Research on Cancer). 2006. *IARC monographs on the evaluation of carcinogenic risks to humans.* Lyon, France: WHO.

Ibrekk, H., and M. G. Morgan. 1987. Graphical communication of uncertain quantities to nontechnical people. *Risk Analysis* 7(4):519-529.

IPCC (Intergovernmental Panel on Climate Change). 2001. *IPCC Third Assessment Report: Climate change 2001 (TAR).* http://www.grida.no/publications/other/ipcc_tar (accessed January 3, 2013).

———. 2007. *IPCC Fourth Assessment Report: Climate change 2007 (AR4).* http://www.grida.no/publications/other/ipcc_tar (accessed January 3, 2013).

Kasperson, R. E., D. Golding, and S. Tuler. 1992. Social distrust as a factor in siting hazardous facilities and communicating risks. *Journal of Social Issues* 48(4):161-187.

Kerr, N. L., and R. S. Tindale. 2004. Group performance and decision making. *Annual Review of Psychology* 55:623-655.

Kloprogge, K., J. van der Sluijs, and A. Wardekker. 2007. Uncertainty communication: Issues and good practice. Version 2.0. http://www.nusap.net/downloads/reports/uncertainty_communication.pdf (accessed September 15, 2008).

Kramer, W., and G. Gigerenzer. 2005. How to confuse with statistics or: The use and misuse of conditional probabilities. *Statistical Science* 20(3):223-230.

Kroll-Smith, J. S., and S. R. Couch. 1991. As if exposure to toxins were not enough: The social and cultural system as a secondary stressor. *Environmental Health Perspectives* 95:61-66.

Krupnick, A., R. Morgenstern, M. Batz, P. Nelson, D. Burtraw, J. Shih, and M. McWilliams. 2006. *Not a sure thing: Making regulatory choices under uncertainty.* Washington, DC: Resources for the Future.

Leiss, W. 1996. Three phases in the evolution of risk communication practice. *Annals of the American Academy of Political and Social Science* 545:85-94.

Lipkus, I. M. 2007. Numeric, verbal, and visual formats of conveying health risks: Suggested best practices and future recommendations. *Medical Decision Making* 27(5):696-713.

Löfstedt, R. 2009. *Risk management in post trust societies.* London: Earthscan.

Moore, D. A., and D. M. Cain. 2007. Overconfidence and underconfidence: When and why people underestimate (and overestimate) the competition. *Organizational Behavior and Human Decision Processes* 103(2):197-213.

Morgan, M. G. 1998. Uncertainty analysis in risk assessment. *Human and Ecological Risk Assessment* 4(1):25-39.

———. 2003. Characterizing and dealing with uncertainty: Insights from the integrated assessment of climate change. *Integrated Assessment* 4(1):46-55.

———. 2009. *Best practice approaches for characterizing, communicating and incorporating scientific uncertainty in climate decision making.* Washington, DC: National Oceanic and Atmospheric Administration.

Morgan, M. G., and M. Henrion. 1990. *Uncertainty: A guide to dealing with uncertainty in quantitative risk and policy analysis.* New York: Cambridge University Press.

Moss, R. H., and S. H. Schneider. 2000. Recommendations to lead authors for more consistent assessment and reporting. In *Guidance papers on the cross cutting issues of the Third Assessment Report of the IOCC*, edited by R. Pachauri, T. Taniguchi, and K. Tanaka. Geneva, Switzerland: IPCC. Pp. 33–51.

NCI (National Cancer Institute). 2011. *Making data talk: A workbook*. Bethesda, MD: National Cancer Institute.

Nelson, D. E., B. W. Hesse, and R. T. Croyle. 2009. *Making data talk: Communicating public health data to the public, policy makers, and the press*. New York: Oxford University Press.

NRC (National Research Council). 1989. *Improving risk communication*. Washington, DC: National Academy Press.

———. 1996. *Understanding risk: Informing decisions in a democratic society*. Washington, DC: National Academy Press.

———. 2008. *Public participation in environmental assessment and decision making*. Washington, DC: The National Academies Press.

———. 2009. *Science and decisions: Advancing risk assessment*. Washington, DC: The National Academies Press.

Peters, E. 2008. Numeracy and the perception and communication of risk. *Annals of the New York Academy of Sciences* 1128(1):1–7.

Peters, E., D. Västfjäll, P. Slovic, C. Mertz, K. Mazzocco, and S. Dickert. 2006. Numeracy and decision making. *Psychological Science* 17(5):407–413.

Peters, E., J. Hibbard, P. Slovic, and N. Dieckmann. 2007. Numeracy skill and the communication, comprehension, and use of risk–benefit information. *Health Affairs* 26(3):741–748.

Peters, E., P. S. Hart, and L. Fraenkel. 2011. Informing patients: The influence of numeracy, framing, and format of side effect information on risk perceptions. *Medical Decision Making* 31(3):432–436.

Peters, R., V. Covello, and D. McCallum. 1997. The determinants of trust and credibility in environmental risk communication: An empirical study. *Risk Analysis* 17(1):43–54.

Pew Research Center for the People and the Press. 2010. *Distrust, discontent, anger and partisan rancor: The people and their government*. Washington, DC: Pew Research Center for the People and the Press.

Renn, O. 1999. A model for an analytic–deliberative process in risk management. *Environmental Science and Technology* 33(18):3049–3055.

———. 2004. The challenge of integrating deliberation and expertise: Participation and discourse in risk management In *Risk analysis and society: An interdisciplinary characterization of the field*, edited by T. McDaniels and M. J. Small. Cambridge: Cambridge University Press.

Renn, O., and D. Levine, eds. 1991. *Credibility and trust in risk communication*. Edited by R. E. Kasperson and P. J. M. Stallen, *Communicating risks to the public*. Netherlands: Kluwer Academic Publishers.

Reynolds, B., and W. S. Matthew. 2005. Crisis and emergency risk communication as an integrative model. *Journal of Health Communication* 10(1):43–55.

Rich, R. C., M. Edelstein, W. K. Hallman, and A. H. Wandersman. 1995. Citizen participation and empowerment: The case of local environmental hazards. *American Journal of Community Psychology* 23(5):657–676.

Russo, J. E., V. H. Medvec, and M. G. Meloy. 1996. The distortion of information during decisions. *Organizational Behavior and Human Decision Processes* 66(1):102–110.

Sandman, P. M. 1986. *Explaining environmental risk*. Washington, DC: EPA.

Shah, P., and E. G. Freedman. 2009. Bar and line graph comprehension: An interaction of top-down and bottom-up processes. *Cognitive Science Society* 3:560–578.

Shah, P., G. M. Maylin, and M. Hegarty. 1999. Graphs as aids to knowledge construction: Signaling techniques for guiding the process of graph comprehension. *Journal of Educational Psychology* 4:690–702.
Singer, E., and P. M. Endreny. 1993. *Reporting on risk: How the mass media portray accidents, diseases, disasters, and other hazards.* New York: Russell Sage Foundation.
Slovic, P. 1993. Perceived risk, trust, and democracy. *Risk Analysis* 13(6):675–682.
———. 2000. *The perception of risk.* London; Sterling, VA: Earthscan Publications.
Slovic, P., and J. Monahan. 1995. Probability, danger, and coercion: A study of risk perception and decision making in mental health law. *Law and Human Behavior* 19:49–65.
Slovic, P., and E. Peters. 2006. Risk perception and affect. *Current Directions in Psychological Science* 15(6):322–325.
Slovic, P., B. Fischhoff, and S. Lichtenstein. 1979. Rating the risks. *Environment* 21(3):14–20.
———. 1981. Perceived risk: Psychological factors and social implications. In *The assessment and perception of risk: A discussion*, edited by F. Warner and D. H. Slater. London: Royal Society. Pp. 17–34.
———. 1982. Facts versus fears: Understanding perceived risk. In *Judgment under uncertainty: Heuristics and biases*, edited by D. Kahneman, P. Slovic and A. Tversky. Cambridge: Cambridge University Press. Pp. 463–492.
Slovic, P., J. Monahan, and D. G. MacGregor. 2000. Violence risk assessment and risk communication: The effects of using actual cases, providing instruction, and employing probability versus frequency formats. *Law and Human Behavior* 24(3):271–296.
Smithson, M., D. V. Budescu, S. Broomell, and H. H. Por. 2011. Never say "not": Impact of negative wording in probability phrases on imprecise probability judgments. Paper read at Proceedings of the Seventh International Symposium on Imprecise Probability: Theories and Applications, July 25–28, Innsbruck, Austria.
Spiegelhalter, D., M. Pearson, and I. Short. 2011. Visualizing uncertainty about the future. *Science* 333(6048):1393–1400.
Stirling, A. 2010. Keep it complex. *Nature* 468(7327):1029–1031.
Stone, E. R., J. F. Yates, and A. M. Parker. 1997. Effects of numerical and graphical displays on professed risk-taking behavior. *Journal of Experimental Psychology: Applied* 3(4):243–256.
Stone, E. R., W. R. Sieck, B. E. Bull, J. F. Yates, S. C. Parks, and C. J. Rush. 2003. Foreground:background salience: Explaining the effects of graphical displays on risk avoidance. *Organizational Behavior and Human Decision Processes* 90(1):19–36.
Sunstein, C. R. 2005. Moral heuristics. *Behavioral and Brain Sciences* 28(4):531–541.
Thompson, K. M., and D. L. Bloom. 2000. Communication of risk assessment information to risk managers. *Journal of Risk Research* 3(4):333–352.
Tversky, A., and D. Kahneman. 1974. Judgment under uncertainty: Heuristics and biases. *Science* 185(4157):1124–1131.
van den Bos, K. 2001. Uncertainty management: The influence of uncertainty salience on reactions to perceived procedural fairness. *Journal of Personality and Social Psychology* 80(6):931–941.
Visschers, V. H. M., and M. Siegrist. 2008. Exploring the triangular relationship between trust, affect, and risk perception: A review of the literature. *Risk Management* 10(3):156–167.
Visschers, V. H. M., R. M. Meertens, W. W. F. Passchier, and N. N. K. De Vries. 2009. Probability information in risk communication: A review of the research literature. *Risk Analysis* 29(2):267–287.
Wallsten, T. S., and D. V. Budescu. 1995. A review of human linguistic probability processing: General principles and empirical evidence. *Knowledge Engineering Review* 10(01):43–62.

Wallsten, T. S., D. V. Budescu, A. Rapoport, R. Zwick, and B. Forsyth. 1986. Measuring the vague meanings of probability terms. *Journal of Experimental Psychology: General* 115(4):348-365.

Wardekker, J. A., J. P. van der Sluijs, P. H. M. Janssen, P. Kloprogge, and A. C. Petersen. 2008. Uncertainty communication in environmental assessments: Views from the Dutch science-policy interface. *Environmental Science and Policy* 11(7):627-641.

Weiss, C. H., E. Singer, and P. M. Endreny. 1988. *Reporting of social science in the national media*. New York: Russell Sage Foundation.

7

Synthesis and Recommendations

Human health risk estimates are a cornerstone of the analyses that the U.S. Environmental Protection Agency (EPA) carries out to help it make its decisions. The uncertainties in those risk estimates have been the subject of much advice from the National Research Council (NRC) and others, many of which have recommended quantitative uncertainty analysis and discussed the use of default assumptions, probabilistic analysis, and the quantitative characterization of uncertainty (NRC, 1983, 1994, 1996, 2009; Presidential/Congressional Commission on Risk Assessment and Risk Management, 1997). EPA has made substantial advances in developing and using analytic techniques for assessing and narrowing the uncertainties in human health risk assessments, and EPA's assessments have evolved and become more complex over the years (GAO, 2006; NRC, 2009). As a recent publication highlights, however, the quantitative or probabilistic uncertainty analyses that are now sometimes conducted can be more extensive than is needed by the decision maker, and might not contribute to EPA's public health and environmental objectives (Goldstein, 2011). NRC reports have also emphasized the need to improve the utility of health risk assessments, and of the uncertainty analyses that are a part of the risk-assessment process (NRC, 1996, 2009).

Through its review of the risk assessment, uncertainty analysis, and decision making at EPA and other public health entities, the committee found that, in general, more emphasis should be placed on the uncertainty in the factors that affect EPA's decisions besides estimates of human health. Uncertainties pervade not only the relationships between hazards and health outcomes, but also other important factors, such as the willingness to pay for improvements in health, the pace of technological change, the importance of recreational areas, and the cost to the private sector of implementing new rules.

As part of a systematic approach to decision making, EPA should plan its analysis of the uncertainty analyses in estimates of health risks and those other factors around the needs of the decision makers beginning at the outset of the decision-making process; in other words, uncertainty analyses should not be an afterthought. Characterizations of health risk estimates, benefits, costs, technological availability, and other factors should reflect those uncertainty analyses. The implications of those different sources of uncertainty are presented in Box 7-1.

The appropriate uncertainty analysis depends, in part, on the type of uncertainty (variability and heterogeneity, model and parameter uncertainty, or deep uncertainty). Statistical analyses are often appropriate for assessing variability and heterogeneity, expert judgments and elicitations work well for model and parameter uncertainty, and robust decision-making approaches and scenario analysis will be needed for decision making in the presence of deep uncertainty.

Good communication among analysts, decision makers, and stakeholders is critical to ensuring a high-quality, comprehensive decision-making process and a high-quality, comprehensive decision. A process that includes such communication will help identify stakeholder concerns and potential uncertainties and to build social trust among the participants in the process. Each EPA decision is unique, and there is no universal best approach or tool for communicating uncertainties. The most appropriate strategy for communicating uncertainty will depend on the context of the decision, the purpose of the communication, the type of uncertainty, and the characteristics of the audience, including the level of technical expertise, personal and group biases, and the level of social trust.

Analyzing and communicating the uncertainty in the various factors that affect EPA's decisions might require specialized expertise (for example, expertise in the analysis of benefits and cost and the uncertainties in each and also in communicating those uncertainties), and some of the necessary skills may be different from those found among EPA's current personnel.

The committee's specific findings and recommendations are presented below.

BOX 7-1
Implications of Uncertainty Analysis for Decision Making

Health Uncertainties
Uncertainty analyses in human health risk estimates can help decision makers to

- evaluate alternative regulatory options;
- assess how credible extreme risk estimates are and how much to rely on them in decision making;
- weigh the marginal decrease in risk against the effort made to reduce it;
- clarify issues within a decision by using scenarios to characterize very different worlds; and
- in the case of scenario analyses for deep uncertainty, identify regulatory solutions that are effective over a broad spectrum of scenarios.

Uncertainties About Technology Availability
Uncertainty analyses in technology availability can help decision makers to

- differentiate between well-established technologies with reasonably well-known costs, and those that have not been used for the purposes at hand; and
- consider which technology may be considered "best practicable" or "best available" by providing information about both the likelihood of success of the unproven technologies the time frame for success, and the effectiveness if successful.

Uncertainties About Cost–Benefit Analyses
Given the highly uncertain estimates of both health benefits and costs, uncertainty analyses in cost–benefit analyses can inform decision makers about

- how difficult it is to differentiate among different potential decisions;
- the disagreement among experts about the way regulation affects the economy, even when using similar models; and
- the ranges and sensitivity of estimates to different variables.

FINDINGS AND RECOMMENDATIONS

Uncertainties in the Characterization of Human Health Risks

Finding 1

Decision documents (such as documents that the technical experts at EPA prepare to explain site-specific decisions) often lack a robust discussion of the uncertainties identified in the extensive health risk assessments prepared by agency scientists. Although decision documents and communications by the agency about its decisions should be succinct, open, and transparent, they should also include information on what uncertainties are present, which uncertainties need to be addressed, and how those uncertainties affected a decision. It should be clear from agency communications that **uncertainty is inherent in science, including the science that informs EPA decisions.**

In addition to contributing to full transparency, providing such information could mitigate attempts to use the existence of uncertainties as a rationale for delayed decision making. A shift in the expectations of those who read the decision documents and other public communications could eventually occur, so that the discussion of uncertainty, even unresolved uncertainty, could eventually come to be considered normal and valuable.

The decision maker will be able to use the analyses most responsibly if the known sources of uncertainties are acknowledged and described, even if the uncertainties are large, poorly described, or not currently resolvable. When uncertainties are so extensive and relevant so as to likely undermine the credibility and quality of a decision, it is the responsibility of staff analysts to advise decision makers as soon as possible in order to give the decision makers a chance to revisit the rulemaking schedule, objectives, and resources, which can require a judgment call.

RECOMMENDATION 1
To better inform the public and decision makers, U.S. Environmental Protection Agency (EPA) decision documents and other communications to the public should systematically

- include information on what uncertainties in the health risk assessment are present and which need to be addressed,
- discuss how the uncertainties affect the decision at hand, and
- include an explicit statement that uncertainty is inherent in science, including the science that informs EPA decisions.

Uncertainty in Other Factors That Influence a Decision

Finding 2

Although EPA decisions have included discussion and consideration of the uncertainties in the health risk assessment, the agency has generally given less attention to uncertainties in other contributors influencing the regulatory decision. Those contributors include economic, and technological factors, as well as other factors that are not easily quantified, such as environmental justice. A major challenge to decision making in the face of uncertainty is the uncertainty in those other factors. Although every uncertainty does not need be analyzed for every decision, particularly when not important to a decision or when specific values are prescribed by other offices (for example, the value of a life that is set by the Office of Management and Budget), methods and processes should be available for when such analyses are appropriate and helpful to a decision maker. In general, this might require a research program to develop methods for this new type of uncertainty analysis, changes in decision documents and other analyses, and a program for research on communicating uncertainties.

RECOMMENDATION 2
The U.S. Environmental Protection Agency should develop methods to systematically describe and account for uncertainties in decision-relevant factors in addition to estimates of health risks—including technological and economic factors—into its decision-making process. When influential to a decision, those new methods should be subject to peer review.

Finding 3

EPA has developed guidance about, and conducted in-depth analyses of, the costs and benefits of major decisions. Some such analyses are conducted because of statutory mandate; others are conducted in response to a series of executive orders from the Office of Management and Budget mandating regulatory review. The analytic tools for cost–benefit analyses in health are well developed, and EPA guidance contains appropriate advice about the conduct of these analyses, including the discussion of some uncertainties. The committee, however, noted a lack of transparency regarding uncertainty analyses in the cost–benefit assessments in some EPA decision documents. While economists or modelers could evaluate the technical documents supporting the decision documents, and the economic models do include uncertainty analyses such as sensitivity analyses, the information about these uncertainties is often arcane, hard to locate, and technically very challenging to non-experts. Those analyses often shape regulatory decisions, and when they are conducted they should be described in ways that are useful and interpretable for the decision maker and stakeholders. The needs of the two audiences—that is, technical and non-expert audiences—differ, but a given set of decision documents and supporting analyses could include descriptions that explain the sources of uncertainties to the non-expert, and link, either electronically or via text, to more detailed descriptions of the economic analyses as appropriate for experts.

RECOMMENDATION 3

Analysts and decision makers should describe in decision documents and other public communications uncertainties in cost–benefit analyses that are conducted, even if not required by statute for decision making, the analyses should be described at levels that are appropriate for technical experts and non-experts.

Finding 4

The role of uncertainty in the costs and benefits and availability and feasibility of control technologies is not well investigated or understood. The evidence base for those factors is not robust. Evaluating case studies of past rulemaking and developing a directed research program on assessing the availability of technology might be first steps toward understanding the robustness of technology feasibility and economic assessments, and the potential for technology innovation.

RECOMMENDATION 4

The U.S. Environmental Protection Agency (EPA) should fund research, conduct research or both, to evaluate the accuracy and predictive capabilities of past assessments of technologies and costs and benefits for rulemaking in order to improve future efforts. This research could be conducted by EPA staff or else by nongovernmental policy analysts who might be less subject to biases. This research should be used as a learning tool for EPA to improve its analytic approaches to assessing technological feasibility.

Finding 5

The committee did not find any specific guidance for assessing the uncertainties in the other factors that affects decision making, such as social factors (for example, environmental justice) and the political context. The committee also did not find examples of systematic consideration of those factors and their uncertainty when exploring the policy implications of strategies to mitigate harms to human health. In response to requirements in statutes or executive orders that require regulations to be based on the open exchange of information and the perspectives of stakeholders, some EPA programs (e.g., Superfund) work to address issues related to public (stakeholder) values and concerns.

Ecological risk assessments[1] have included contingent valuation to help inform policy development. Economists have similarly explored the values people hold regarding specific health outcomes for purposes of resource allocation or clinical guideline development. More research is needed into methods to appropriately characterize the uncertainty in those other factors, and to communicate that uncertainty to decision makers and the public.

RECOMMENDATION 5
The U.S. Environmental Protection Agency should continue to work with stakeholders, particularly the general public, in efforts to identify their values and concerns in order to determine which uncertainties in other factors, along with those in the health risk assessment, should be analyzed, factored into the decision-making process, and communicated.

Finding 6
The nature of stakeholder participation and input to decision makers depends on the type of stakeholder. The regulated industry, local business communities, and environmental activists (including at the local level if they exist) are more likely to be proactively engaged in providing input on pending regulations. The general public, without encouragement or assistance from EPA (or local environmental regulatory departments), is less likely to participate effectively or at all in such activities. One means to bridge the gap in understanding the values of the public is a formal research program.

RECOMMENDATION 6
The U.S. Environmental Protection Agency should fund or conduct methodological research on ways to measure public values. This could allow decision makers to systematically assess and better explain the role that public sentiment and other factors that are difficult to quantify play in the decision-making process.

---

[1] Ecological risk assessment is a "process that evaluates the likelihood that adverse ecological effects may occur or are occurring as a result of exposure to one or more stressors" (http://www.epa.gov/raf/publications/pdfs/ECOTXTBX.PDF [accessed January 16, 2013]).

Framework for Incorporating Uncertainty in Decision Making

Finding 7

Uncertainty analysis must be designed on a case-by-case basis. The choice of uncertainty analysis depends on the context of the decision, including the nature or type of uncertainty (that is, heterogeneity and variability, model and parameter uncertainty, or deep uncertainty), and the factors that are considered in the decision (that is, health risk, technological and economic factors, as well as other issues such as public sentiment and the political context), as well as the data that are available. Most environmental problems will require the use of multiple approaches to uncertainty analysis. For example, most environmental decisions will involve variability and heterogeneity as well as model and parameter uncertainty. As a result a mix of statistical analyses and expert judgments will be needed.

A sensible, decision-driven, and resource-responsible approach to uncertainty analyses is needed. Iterative and deliberative problem formulation and planning using a systematic framework for the decision-making process will help ensure that the nature and extent of uncertainty analysis to be included in risk characterizations is appropriate to the decision context, and that decision makers are provided a view of uncertainty that is of maximum value to the decision under consideration. Involvement of decision makers and stakeholders in the planning and scoping of uncertainty analyses during the initial, problem formulation phase of will help ensure that the goals of the uncertainty analysis are consistent with the needs of the decision makers, and will help define analytic endpoints, and identify population subgroups and heterogeneity, and other uncertainties.

The quantitative uncertainty analyses that are conducted do not always influence a decision and, therefore, do not always contribute to protection of public health. The committee does not intend to imply that complex uncertainty analyses have no role to play in supporting EPA decision making; rather the committee believes that such work should only undertaken when it is important and relevant to a given decision. Whether further quantitative uncertainty analysis is needed depends on the ability of these analyses to affect the environmental decision at hand. One way to gauge this is to inquire, whether perfect information would be able to change the decision, for example, whether knowing the exact dose–response function would change the regulatory regime. Clearly, if the environmental decision would stay the same for all states of information and analysis results, then it is not worth conducting the analysis.

RECOMMENDATION 7

Although some analysis and description of uncertainty is always important, how many and what types of uncertainty analyses are carried out should depend on the specific decision problem at hand. The effort to analyze specific uncertainties through probabilistic risk assessment or quantitative uncertainty analysis should be guided by the ability of those analyses to affect the environmental decision.

Communication

Finding 8

A structured format for the public communication of the basis of EPA's decisions would facilitate transparency and subsequent work with stakeholders, particularly community members. Consistent with findings and analyses in each rulemaking, EPA decision documents should make clear that the identified uncertainties are in line with reasonable expectations presented in EPA guidelines and other sources. This practice would facilitate the goals of the first recommendation of the committee in this report—that *EPA decision documents should make clear that uncertainty is inherent in agency risk assessments*. The committee intends that the recommendations in this report support full discussion of the difficulties of decision making, including and possibly particularly when social factors (such as environmental justice and public values) and political context play a large role.

RECOMMENDATION 8.1

U.S. Environmental Protection Agency senior managers should be transparent in communicating the basis of its decisions, including the extent to which uncertainty may have influenced decisions.

RECOMMENDATION 8.2

U.S. Environmental Protection Agency decision documents and communications to the public should include a discussion of which uncertainties are and are not reducible in the near term. The implications of each to policy making should be provided in other communication documents when it might be useful for readers.

Finding 9

Given that decision makers vary in their technical backgrounds and experience with highly mathematical depictions of uncertainty, a variety of communication tools should be developed. The ability of the public to assimilate the depictions of uncertainty and risk estimates is even more diverse. The public increasingly wants, and deserves, the opportunity to understand the decisions of appointed officials in order to manage their own risk, and to hold decision makers accountable. With respect to *which* uncertainties or aspects of uncertainties to communicate, attention should be paid to the relevance to the audience of the uncertainties, so that the uncertainty information is meaningful to the decision-making process and the audience(s). Those efforts should include different types of decisions and include communication of uncertainty to decision makers and to stakeholders and other interested parties.

RECOMMENDATION 9.1
The U.S. Environmental Protection Agency (EPA), alone or in collaboration with other relevant agencies, should fund or conduct research on communication of uncertainties for different types of decisions and to different audiences, develop a compilation of best practices, and systematically evaluate their communications.

RECOMMENDATION 9.2
As part of an initiative evaluating uncertainties in public sentiment and communication, U.S. Environmental Protection Agency senior managers should assess agency expertise in the social and behavioral sciences (for example, communication, decision analysis, and economics), and ensure it is adequate to implement the recommendations in this report.

In summary, the committee was impressed by the technical advances in uncertainty analysis used by EPA scientists in support of EPA's human health risk assessments, which form the foundation of all EPA decisions. The committee believes that EPA can lead the development of uncertainty analyses in economics and technological assessment that are used for regulatory purposes, as well as how to characterize and account for public sentiment and political context. That leadership will require a targeted research program, as well as disciplined attention to how those uncertainties are described and communicated to a variety of audiences, including the role that uncertainties have played in a decision.

REFERENCES

GAO (Government Accountability Office). 2006. *Human health risk assessment: EPA has taken steps to strengthen its process, but improvements needed in planning, data development, and training*. Washington, DC: GAO.
Goldstein, B. D. 2011. Risk assessment of environmental chemicals: If it ain't broke.... *Risk Analysis* 31(9):1356–1362.
NRC (National Research Council). 1983. *Risk assessment in the federal government: Managing the process*. Washington, DC: National Academy Press.

———. 1994. *Science and judgment in risk assessment*. Washington, DC: National Academy Press.
———. 1996. *Understanding risk: Informing decisions in a democratic society*. Washington, DC: National Academy Press.
———. 2009. *Science and decisions: Advancing risk assessment*. Washington, DC: The National Academies Press.
Presidential/Congressional Commission on Risk Assessment and Risk Management. 1997. *Risk assessment and risk management in regulatory decision-making. Final report. Volume 1*. Washington, DC.

Appendix A

Approaches to Accounting for Uncertainty

This appendix discusses a number of different approaches available for analyzing uncertainty and considering uncertainty in decisions. It provides greater detail on the approaches to analyzing or accounting for uncertainties in decisions that are discussed in Chapter 5. Most environmental problems require the use of multiple approaches to uncertainty analysis. For example, most environmental decisions involve variability and heterogeneity as well as model and parameter uncertainty. As a result, it is necessary to apply a mix of statistical analyses and expert judgments.

HEALTH ONLY

When assessing human health risks, the main uncertainties arise in projecting exposures and health effects in the baseline case—that is, absent a change in a risk management strategy—and in projecting the effects of a given management intervention (for example, a proposed regulatory action, such as the implementation of an emission standard). Variability and heterogeneity can occur because of variability both in exposures and in sensitivity to the exposure among subgroups of the population, and existing data may be inadequate to accurately characterize the underlying heterogeneity. Further uncertainty can arise when using models that combine multiple health effects into a single outcome measure.

THE USE OF DEFAULTS

Many U.S. Environmental Protection Agency (EPA) decisions consider only health factors. EPA's primary general approach for considering uncertainty in this class of problems has been to use safety or default adjustment factors. (See Chapter 2 for a discussion of the use of default adjustment factors, or defaults.) The decision rule for these approaches is to set a standard or regulation that is highly protective by applying defaults. These approaches are health protective in nature, widely used, and sometimes embodied in statutes. As discussed in *Science and Decisions* (NRC, 2009), many of the defaults that EPA uses were developed on a scientific basis and can be adequate and acceptable to use in some risk assessments. For example, in instances when there is not adequate information or when the potential uncertainties are such that the use of defaults compared to quantitative uncertainty analyses is unlikely to affect a decision, defaults can be used.

One of the main objections that decision analysts have to using default factors is that they incorporate implicit judgments by analysts or scientists who do not make the regulatory decision. Furthermore, those judgments and their implications are not always independent and are not always explained to decision makers, which makes it difficult for the decision makers to properly interpret the assessment in the context of other factors.

Using health-protective (called *conservative*) analytic or default approaches to account for multiple uncertainties can result in an overestimation of health risks and a level of precaution in excess of one based on expected values (Nichols and Zeckhauser, 1988; Viscusi et al., 1997). With this happens—a situation sometimes referred to as compounding conservatism—the precaution level for each individual analysis might be such that the marginal cost of precaution equals or slightly outweighs the marginal health benefit, but when multiple analyses use that level of precaution and are combined, the precaution level becomes such that the overall marginal cost far exceeds the overall marginal benefit. It is unclear, however, how extensive that problem is in reality, and, as discussed by the Government Accountability Office, EPA has taken steps to improve such analyses and avoid some of the problems of compounding conservatism (GAO, 2006). Cullen (1994) evaluated the effects of potential compounding conservatism and found that "there exist cases in which conservatism compounds dramatically, as well as those for which the effect is less notable" (p. 392).

To the extent that the probability distribution function is flat and wide (that is, it has "fat tails"; see Farber, 2007, for discussion) rather than being tall and single-peaked, the safety factor will be high relative to the expected value. Conversely, if the probabilities of adverse outcomes are very low, the cost may be low. Basing decisions on values in the left or right tail of a distribution rather than on the mean of the distribution will typically result in a suboptimal allocation of scarce resources, even in the presence of uncertainty. Such decisions can also lead to opportunity costs in the form of wasted resources as well as cynicism about the benefits of regulation.

VARIABILITY AND HETEROGENEITY

Variability and heterogeneity (that is, randomness) are often seen in environmental conditions, exposure levels, and the susceptibility of individuals. When there is information about this randomness with which to conduct statistical analyses, sometimes including extreme value analyses, it is appropriate to use such analysis to generate distributions. Randomness typically occurs in an equation's error term, in parameters describing the relationships, or both. There may be structural differences in the relationships among population subgroups (Table 5-1, first row, first column). There are important considerations that the statistician or epidemiologist should address when designing appropriate statistical models and procedures, such as the choice of the assumptions underlying a specific random process (for example, a binomial or Poisson process) or the assumptions of independent sampling. Those issues are well covered by textbooks on statistics.

The use of statistics-based probabilistic risk analysis is an alternative to using safety or uncertainty factors. Probabilistic risk analysis uses probability distributions to quantify uncertainty at each step of the risk assessment. For example, a probabilistic risk analysis would quantify the uncertainties about a dose–response relationship for exposure to fine particulates by using epidemiological studies to construct a probability distribution around the slope of the dose–response function.

The typical decision rule in a probabilistic risk analysis is to select a standard or regulatory option that satisfies a specific probability criterion. Thus probabilistic risk analysis and decision analysis[1] are well suited to deal with the challenge of ensuring an acceptable margin of safety. For example, if a distribution describes the lifetime cancer risk of a maximally exposed individual, analysts can provide the decision maker with the probability that cancer will occur in such an individual—for example, 1 chance in 100, 1 in 10,000, 1 in 100,000, 1 in 1 million, or 1 in 10 million—for a variety of different regulatory options. The decision maker can then decide which of those probabilities is acceptable and choose a regulatory option that reduces the probability to that level. The advantage of using such analytic

---

[1] Decision analysis uses a systematic approach to make decisions in the face of uncertainty. In contrast, probabilistic risk analysis estimates risks, in this case human health risks, using probabilistic statistical methods.

approaches and such a decision rule is that it characterizes the whole range of the probability distribution of outcomes and there is an explicit choice of the level of risk that is tolerable. This enables the decision maker to explain the decision process, including his or her values, quite clearly. For example, the estimated health risks associated with multiple concentrations of a chemical could be presented, and a decision could then be made as to what health risk is acceptable. As discussed in Chapter 2, such an approach was used in the National Research Council's (NRC's) report *Arsenic in Drinking Water: Update 2001* (NRC, 2001). That report presented the estimated risks of bladder and lung cancer at 3, 5, 10, and 20 ppb of arsenic in drinking water, leaving it for EPA to decide, taking other factors such as costs into consideration, what cancer risk was acceptable and, therefore, what level of arsenic should be allowed in drinking water.

Sometimes the standard tools and techniques for assessing uncertainty can be difficult to use when faced with the uncertainty depicted by the first two rows of Table 5-1 (that is, variability and heterogeneity, or model and parameter uncertainty). Probabilistic risk analysis often makes assumptions about the underlying probability distributions that describe the risk (for example, whether a particular distribution is normal or log-normal). The possibility of nonregular probability distributions makes performing a probabilistic risk analysis more difficult, especially when the underlying distribution for a parameter has what is termed a "fat tail." Distributions of extreme events can have fat tails (for discussion, see Farber, 2007), and under those circumstances the "fat tail" has to be modeled.

For example, in the aftermath of Hurricane Katrina's devastation of New Orleans, Southwell and von Winterfeldt (Southwell and von Winterfeldt, 2008) noted that when the U.S. Army Corps of Engineers was designing and building levees and floodwalls in the 1970s and 1980s, the corps estimated that a Category 4 hurricane would occur once in 100 years in New Orleans (U.S. Army Corps of Engineers, 1984). That estimate was made despite the fact that two hurricanes rated Category 3 or higher had hit Louisiana not far from New Orleans in the 20 years prior to that estimate being made: In 1965, while still in the Gulf of Mexico, Hurricane Betsy peaked out at a force that was rated just below a Category 5 storm, and when its eye made landfall southeast of New Orleans at Grand Isle, Louisiana, Betsy was rated as a Category 3; and in 1969 Hurricane Camille made landfall in Mississippi as a Category 5 storm. The fact that Hurricane Gustav, a Category 3 hurricane, hit New Orleans only 3 years after the near-Category 4 Hurricane Katrina raised more questions about the 100-year estimate for the occurrence of Category 4 hurricanes and about whether the statistical methods used for the prediction were appropriate.

Statisticians (Cooke et al., 2011) and decision theorists (Bier et al., 1999) have proposed the use of methods that emphasize extremes in order to adjust for a fat-tail distribution when dealing with certain rare but potentially catastrophic events. However, such an approach assumes that the inaccurate predictions are the result of a stable, underlying distribution that has been somehow mis-specified. But it is certainly possible that the inaccurate predictions are instead the result of a structural change and, therefore, that the probability distributions drawn from the historical record are no longer relevant. For example, the levels of protection provided by the levees and floodwalls around New Orleans decreased over time because of natural and man-made changes, and those decreases were not accounted for in the models. As a practical matter, however, it may not be possible to distinguish between the two sources of error.

MODEL AND PARAMETER UNCERTAINTY

Expert elicitations are often used when dealing with uncertainty about what statistical model to use and which parameters to use in the model (second row, first column in Table 5-1). An expert elicitation is "a formal systematic process to obtain quantitative judgments on scientific questions (to the exclusion of personal or social values and preferences)" (EPA, 2011, p. 23).[2] For example, the U.S. Food and Drug Administration and its Food Safety and Inspection Service used expert elicitation to rank foods according to their ability to support the growth of *Listeria monocytogenes* (see Chapter 4 for discussion) (FDA, 2003).

Expert elicitation and statistical analysis are not mutually exclusive, and a decision maker may choose to use both methods. Expert elicitation processes have been designed and used to quantify model and parameter uncertainties (Hora, 2007). For example, EPA had a formal expert elicitation conducted to incorporate "expert judgments into uncertainty analyses for the benefits of air pollution rules" concerning particles less than 2.5 micrometers in diameter using "carefully structured questions about the nature and magnitude of the relationship between changes in annual average $PM_{2.5}$ and annual, adult, all-cause mortality in the U.S." (Industrial Economics Incorporated, 2006, p. ii). Expert elicitations can also be used in combination with Bayesian updating models. For instance, elicitations can be used to revise or set initial probability distributions rather than using data to estimate posterior distributions (that is, to reflect the state of existing knowledge before incorporating new evidence through a Bayesian analysis). Although many issues are still disputed—such as whether to elicit experts individually or in groups, how to aggregate different opinions when experts are elicited individually, and how to combine expert opinion with the results of statistical analysis if both techniques are used (Leal et

---

[2] Expert elicitation is described in detail elsewhere (see EPA, 2011; Slottje et al., 2008).

al., 2007)—the process of using experts and Bayesian updating is becoming more widely used (Choy et al., 2009; Kuhnert et al., 2010). Such an approach has been applied in various areas, including health care resource allocation (Griffin et al., 2008), conservation science (Martin et al., 2012; O'Leary et al., 2008), ecological models (Kuhnert, 2011), coral reef protection (Bouma et al., 2011), and animal diseases (Garabed et al., 2008).

The uncertainty that results from analytical model uncertainty—that is, not knowing what statistical model should be used to estimate relationships such as the dose–response relationship—is difficult to quantify, but in some cases it can have a more pronounced effect on estimates of health risk than parameter uncertainty. Rhomberg (2000), for example, showed the wide range or risk estimates that could result from applying different models to the results of studies of trichloroethylene exposures in mice. Similarly, the choice of which statistical model to use to extrapolate from a high-exposure occupational study to low-dose exposures can have an order-of-magnitude effect on estimates of health risk. As discussed by NRC (2006), to explore the effects of model choice, in its risk assessment EPA estimated the point of departure for determining an acceptable dose (in this case, one that produces a 1 percent change in the risk of cancer, or ED_{01}) using data from a dioxin occupational exposure study (Steenland et al., 2001) but using two different statistical models.[3] The point of departure estimated using a power statistical model was 1.38 ng/kg, but the corresponding value estimated with a linear model was 18.6 ng/kg (NRC, 2006). The point of departure was more than an order of magnitude higher when the extrapolation to a low dose was done with linear model rather than a power model; this implies that choosing a linear extrapolation rather than a power extrapolation could result in a regulatory standard more than an order of magnitude more stringent. Disagreements about which model is appropriate for low-dose extrapolations of the cancer risks of dioxin have resulted in extensive delays in finalizing the dioxin health risk assessment. Although expert elicitation might be able to guide the decision of which model is most appropriate, there are only a handful of examples of using expert elicitation processes for this purpose in the environmental setting. One example is the work of Ye et al. (2008), who used expert elicitation to weight five models developed for the Death Valley regional flow system.

Developing models to deal with multiple datasets is an active area of research. For example, the National Cancer Institute funds the Cancer Intervention and Surveillance Modeling Network (CISNET), whose goal is to develop tools to assist in synthesizing cancer-related evidence (NCI, 2012).

---

[3] The piecewise linear model function was e^{bx}. The power model function was x/background (EPA, 2003; NRC, 2006).

In other instances expert committees have recommended that, in the absence of a biological rationale for choosing one particular model—such as a well-established mode or mechanism of action that indicates how a low dose of a chemical induces and initiates cancer—the choice of a model should be driven by the fit of the existing data to a statistical model and the biological plausibility of the model. For example, *Arsenic and Drinking Water: 2001 Update* (NRC, 2001) found the available information about how arsenic might cause cancer (that is, the mode-of-action data) "insufficient to guide the selection of a specific dose–response model," and recommended an "additive Poisson model with a linear term in dose [because it] is a biologically plausible model that provides a satisfactory fit to the epidemiological data and represents a reasonable model choice for use in the arsenic risk assessment" (p. 209). That report also assessed the impact of model choice on risk estimates in order to provide information on whether model choice is an important source of uncertainty.

If there is insufficient time, if there is insufficient consensus information for expert elicitation and subjective model weighting or averaging to estimate model uncertainty, or if such uncertainty analyses are not required given the context of the decision, then one can choose to use a structured system of model defaults and criteria for departures from those defaults. *Science and Decisions* (NRC, 2009) highlights the delays that have occurred because of disagreements about the "adequacy of the data to support a default or an alternative approach" (p. 7), and recommends that EPA "describe specific criteria that need to be addressed for the use of alternatives to each particular default assumption" (p. 8).

TECHNOLOGY AVAILABILITY

As discussed in Chapters 1 and 3, in addition to considering estimates of health risks, some EPA decisions consider the availability—either current availability or availability that is expected in the foreseeable future—of technology necessary to achieve a desired exposure reduction and health outcome (second column in Table 5-1). In such cases, two questions arise in addition to the questions related to estimates of health risks. First, which technologies are available or likely to be available soon that can achieve the desired reductions in risk? Second, if several technologies might achieve the health objectives, which one or ones are most suitable? The choice of technology can depend on several factors, including current versus future availability, effectiveness in decreasing exposures and improving health outcomes, and the cost of investments in the new technology, irrespective of how these costs are borne. As discussed in Chapter 3, the availability of new technology is not independent of rulemaking. Rulemaking can, in effect, create a market for technologies. If entrepreneurs believe there will be a market for a new product, they will be more likely to invest in the research and development of such a product. Questions about whether those potential markets will spur the development of new technologies by the time a regulation is implemented add to the uncertainty concerning the technology. Expert elicitations and expert judgments can provide much of the needed information concerning technology availability.

In answering the first question—which appropriate technologies are available or soon will be—a key issue is the uncertainty about the likelihood that a relatively new technology can be successfully deployed and, if it is successfully deployed, how well it will perform. In discussing how to ameliorate the effects of the growing amounts of carbon dioxide in the atmosphere, for example, one might ask which carbon sequestration technologies can be successfully deployed and, of those, which will have the best performance. Decision tree analysis, which is one of several analytic techniques used in the field of decision analysis (Clemen, 1998; Raiffa, 1968), is useful for answering such questions. In this case the branches of the decision tree would represent a particular technology being available or not available at some specified time period in the future, with probabilities attached to each outcome.

Suppose that the decision involves mutually exclusive choices among technologies, not all of which are available at the time the decision is being made. The decision tree lays out initial decision options (technologies), and for those technologies that do not yet exist the option is followed by nodes that represent the chance for success or failure in implementing the technology. Success and, separately, failure are each followed by additional decisions that represent, for example, adjustments to the technology that are likely to occur in response to the initial success or failure, which are followed by final chance nodes,[4] such as indicators of actual performance and the utility or gain measured in some other way that is associated with an actual performance level. One technology may be more promising than another, in which case the actual performance level will be higher—as will the associated utility. EPA's estimates of technological advances are further complicated because although the agency establishes standards and conducts analysis of the costs and benefits of technologies, other sectors (for example, industry) will develop and use the particular technology in question.

When technologies vary on multiple dimensions—for example, cost, performance, and reliability—an analysis of the trade-offs between these dimensions is needed. Such analyses are discussed below.

---

[4] A chance node is an event or point in a decision tree where a degree of uncertainty exists.

COST–BENEFIT COMPARISONS

Cost–benefit trade-offs can be analyzed by cost-effectiveness or cost–benefit analysis (referred to as economic factors in Figure 1-2 and defined and discussed at length in Chapter 3; see also third row in Table 5-1). Cost-effectiveness analysis is used much more widely than cost–benefit analysis for decisions involving personal and public health (Gold, 1996; Sloan and Hsieh, 2012). By contrast, cost–benefit analysis is much more widely used for business[5] and is the focus of EPA's environmental applications, although the Office of Management and Budget has recommended that the EPA also use cost-effectiveness analyses (OMB, 2012). A major advantage of cost–benefit analysis is that nonhealth benefits can be included along with health benefits in the benefit calculation.

A number of studies have evaluated the quality of cost–benefit analyses conducted to support regulatory decisions, including environmental decisions. Agency cost–benefit analyses have been criticized for not consistently providing a range of total benefits and costs, and information on net benefits (Ellig and McLaughlin, 2012; Hahn and Dudley, 2004; Hahn and Tetlock, 2007). A number of studies have assessed the extent to which the outcome of those analyses affected the regulatory decision for which they were performed, often finding that it was not clear exactly how the analyses were considered in making the decision (Hahn and Dudley, 2007; Hahn and Tetlock, 2008). It does seem, however, that such analyses are becoming more widely used in regulatory decisions (Ellig and McLaughlin, 2012).

As discussed in Chapter 3, one uncertainty when evaluating costs and benefits is which costs and benefits to include in the analyses. It is important to determine during the problem-formulation phase at the start of the decision-making process which costs and benefits should be included. For example, there needs to be a decision on whether the economic analysis should include employment loss secondary to an environmental disaster. Here the analyst is reliant on the decision maker's input in defining program objectives and limiting the scope of the analysis, a situation that highlights the importance of having both the decision makers and the analysts involved in this first phase.

The decision rule in cost–benefit analyses is to select the regulatory option with the largest expected net social benefit. As with other decision rules, the decision maker is using predicted estimates for the future; the decisions, therefore, are based on estimates of future values. These estimates reflect the underlying probability distributions of potential costs and benefits, and the approaches require an explicit consideration of the underlying

---

[5] Cost–benefit analyses for business applications typically are not made public and are conducted to provide information to maximize profits. Environmental cost–benefit analyses take social benefits and costs into account.

probabilities of various outcomes and of the costs and benefit or utilities associated with each outcome.

When the outcomes of decisions are reasonably well defined and the underlying probability distributions are reasonably well characterized, these techniques work reasonably well. Problems may arise in practice because, for example, the heterogeneity of dose response is not adequately characterized, issues arise concerning whether existing data can be generalized to the decision problem, or there is disagreement about how to value an endpoint. For most of these issues, additional research may be the answer. The value-of-information approach described below provides a formal approach for determining the benefit of further research.

There are also circumstances in which the decision rules do not work well. First, there may be substantial investment costs associated with certain policy options. If for some reason the investment returns are much less than expected, the large cost of the investment must still be paid. The widely used analytic decision-making tools are quite useful when it is possible to make midcourse corrections in response to new information gained with experience. If the consequences of implementing a policy are irreversible or very costly to reverse, however, basing decisions on expected values may be highly inadvisable, and it may be more appropriate to give more weight in the decision to the nonfavorable (and irreversible) outcomes, which is a health-protective approach.

The *decision rule* in multiattribute utility analysis is to select the regulatory option that maximizes expected utility. It has been used in various health applications (Feeny et al., 2002; Orme et al., 2007), and multiattribute utility analysis and other multicriteria decision analysis tools have also been applied to decisions related to environmental issues (for a review, see Kiker et al., 2005). For example, Merkhofer and Keeney (1987) applied multiattribute utility analysis to help the Department of Energy determine a storage location for nuclear waste. It has also been used for decisions related to the management of the spruce budworm in Canadian forests (Levy et al., 2000) and the selection of a management approach for the Missouri River (Prato, 2003).

In multiattribute utility analysis, utility is a function of each attribute taken individually as well as in interaction with one another (Clemen and Lacke, 2001; Keeney and Raiffa, 1976; Morris et al., 2007). In the context of public policy decision making, attributes, or the probability of various attributes, are associated with each policy option. For instance, cleaning up a site may have several types of payoffs, such as improving various health outcomes and, if factors other than health are considered in the decision, fostering the development of new approaches for clean up, promoting local economic development, and providing recreational opportunities.

Multiattribute utility analysis is particularly useful when valid and reliable utility-of-attribute weights are available from an existing source, as is the case for a number of health applications. Such utility weights may not be as readily available for environmental decision-making applications, in which case the weights would have to be derived as part of the analytic process. Such weights typically vary among individuals and groups, so there is the question of whose weights to use—those of the decision maker, of the stakeholders, or of the members of the public at large—and how best to elicit those weights from the various groups.

The weights assigned might vary among different people and over time. For example, the weights for additional lives saved might vary for different people over time; the value of years of life saved would be different for, say, a 90-year-old than for a newborn baby. At the individual patient level, medical decision makers attempt to maximize expected utility for the patient, rather than simply maximizing the patient's expected life-years. At the population level, medical decision-making guidelines use quality-adjusted life-years (QALYs), computed as a population average, to weight lives saved (Gold et al., 2002).

A number of characteristics of multiattribute utility analysis make it useful to environmental decision makers. First, it can explicitly address the uncertainties of the regulatory problem, including uncertainties concerning the performance of the technologies (discussed on page 238), the risks and the risk reduction that is achievable, and other factors important in the evaluation of technologies. Second, it uses judgments of decision makers to quantify reasonable and defensible trade-offs among the impacts of technology options, which can be informed by—but are not necessarily equal to—those obtained from market studies and surveys. Third, it can account for risk aversion in length of life or other outcomes—that is, it uses a nonlinear utility function over outcomes. In other words, the analyst working with the decision maker can define the utilities for the analysis in a way that best reflects the decision maker's preferences or the preferences of others who have an important voice in the decision. Even such concepts as equity can be assigned a utility value. Although cost–benefit analyses can use nonlinear utility functions, in practice a cost–benefit analysis typically employs a linear utility function. Some research has explored ways to include inequality and inequity in benefits analysis (Levy et al., 2006).

One problem with the use of aggregate analysis—and multiattribute utility analysis in particular—is that the output numbers can be difficult to interpret. Many judgments are typically buried within those numbers, such as the relative weights given to different parameters. The relative weights should be explicitly stated when using such analyses, but it is a challenge to figure out how to display that and other embedded information.

DEEP UNCERTAINTY

Deep uncertainty is uniquely challenging for decision makers. The traditional analytic approaches discussed above, which focus on the probability of certain consequences resulting from different regulatory decisions, cannot typically be used when too little is known about—or there is substantial disagreement concerning—variability, heterogeneity, the appropriate model for the data, or the parameters that should be input into a model (Lempert, 2002). Cox (2012) emphasizes the shortcomings of those traditional methods in cases when there is uncertainty or disagreement about (1) what regulatory alternatives are available; (2) the full range of possible consequences; (3) the correct model for a consequence, given a particular decision; and (4) the values and preferences that should be used to evaluate potential consequences, such as how much value should be given to future generations having a certain resource. In other words, those traditional analytical approaches are not particularly useful when deep uncertainty is pervasive, and judgment calls are necessary.

Robust management strategies, including adaptive management strategies, can be useful when deep uncertainty is present (Cox, 2012; Flüeler, 2001; Lempert, 2002; Lempert and Collins, 2007). Adaptive management strategies characterize uncertainty by using multiple representations of the future rather than a single set of probability distributions, as in optimum expected utility analysis (Lempert and Collins, 2007). An adaptive strategy might give up some level of "optimal performance for less sensitivity to violated assumptions," or be designed to "perform reasonably well over a wide range of plausible futures" (Lempert and Collins, 2007, p. 1016), or it might leave multiple options open. In other words, with an adaptive strategy, a decision maker might "choose strategies that can be modified to achieve better performance as one learns more about the issues at hand and how the future is unfolding" rather than choosing a strategy on the basis of a certain risk estimate (CCSP, 2009). Diversification of financial portfolios is one example of an adaptive strategy (CCSP, 2009). Another important characteristic of such strategies is that they are likely to be adaptable once additional information has been received (Lempert et al., 2003). Two tools that can provide information that will help decision makers develop adaptive strategies—scenarios and value-of-information assessments—are discussed below.

THE USE OF SCENARIOS

One of the methods for making decisions in the face of deep uncertainty is the use of scenarios that specify alternative outcomes based on alternative assumptions about the future (Lempert, 2002). As discussed by Jarke et al. (1998), a scenario describes the set of events that could, within reason, take place. Developing scenarios stimulates participants to identify what situations might occur, the assumptions involved in those situations, what opportunities and risks are associated with the different situations, and what actions could be taken under the different situations.

Rather than asking what is most likely to occur, as traditional analytic approaches do, scenarios explore "questions of what are the consequences and most appropriate responses under different circumstances" (Duinker and Greig, 2007, p. 209). In other words, traditional analytical approaches seek to estimate the likelihood of an event or consequence, while scenarios serve to replace unknowns with conceptually feasible but hypothetical events. The scenarios can span the range of possible future worlds, and, given the set of scenarios, analysts can use traditional methods to assess the likely risks and impacts under each scenario. Instead of maximizing expected value or expected utility, as is done when uncertainty can be quantified, the goal of a scenario analysis is to find a solution that performs well compared to alternative options under a number of dissimilar, albeit plausible, scenarios depicting the future. As discussed in Chapter 4, scenarios have been used to assess the expected results of different regulatory options for controlling bovine spongiform encephalopathy and *Listeria monocytogenes*.

In the scenario approach, attaching probabilities to individual scenarios is discouraged. Instead, the goal is to find regulatory solutions that assure that the risks will be contained even if the worst-case scenario comes true. However, even in a worst-case scenario the option selected may not be the most protective one because protection comes at a cost and complete containment may be wasteful. Some departments, such as the Department of Defense, have moved away from examining the worst-case scenario and focus instead on the more likely scenarios. By examining a number of different scenarios in human health risk assessments, including scenarios using defaults, EPA could examine the effects of the different scenarios, and risk management decision makers could choose the scenario that produces their desired level of precaution for the decision context.

The scenarios are constructed as part of the process of evaluating uncertainty. When there are disputed values—including when stakeholders disagree about what a value should be—the scenarios examined can include ones that incorporate disputed values, thus incorporating into the scenarios the uncertainty that surfaced during the deliberative processes with stakeholders.

Computers make it possible to evaluate a large number of scenarios. This general approach has been used for assessing long-term global economic growth (Lempert et al., 2003), public and private investment in hydrogen and fuel-cell technologies (Mahnovski, 2007), managing the risk of a catastrophic event involving pollution of a pristine lake (Lempert et al., 2006), and the potential effects of various climate-change assessments (Kandlikar et al., 2005).

One of the limitations of using scenarios as a robust approach is that it typically leads to relatively conservative strategies because if a risk-management strategy is to be robust, it has to perform reasonably well even for worst-case scenarios. The approach also requires large computational capabilities compared with more traditional decision analysis methodologies, and it requires the ability to determine the potential consequences of the different scenarios (Lempert et al., 2006). Furthermore, in some cases, depending on the nature of the decision and the evidence that is available, there are no robust solutions, that is, "no amount of effort will suggest strategies that perform reasonably well across all or most plausible states of the world" (Lempert et al., 2006, p. 238). In other words, scenarios and robust decision making cannot solve every problem.

WHEN IS UNCERTAINTY DEEP UNCERTAINTY?

Although in decision making it is useful to recognize when there is deep uncertainty, it is also important to remember that the line between deep uncertainty and other types of uncertainty is not always absolute and can change over time. For example, when dealing with nuclear waste management, the time horizons are on the order of 10,000 to 100,000 years, and it is not possible to know what will happen over that time frame, especially given the possibility of long-term geological issues, seismology issues, volcanic activity, climate change, and future human intrusions (DOE, 2002). As time goes on, however, some of those uncertainties might become more or less deep.

There is no operational definition for when a lack of consensus about the appropriate models for a particular decision-making problem becomes a case of deep uncertainty in which robust decision-making tools would be helpful. A solution that may work in some cases would be to use both traditional decision-making analysis techniques (such as expected utility) and robust decision making that is based on scenarios of possible future states of the world. However, all uncertainty analysis—and deep uncertainty analysis in particular—is costly. So as part of the initial problem-formulation phase of decision making, one should consider whether such uncertainty analyses are likely to affect the decision and thus be worth including in the process. Although this will not always identify the best method, in some instances it can eliminate some options from consideration.

REFERENCES

Bier, V. M., Y. Y. Haimes, J. H. Lambert, N. C. Matalas, and R. Zimmerman. 1999. A survey of approaches for assessing and managing the risk of extremes. *Risk Analysis* 19(1):83–94.

Bouma, J. A., O. Kuik, and A. G. Dekker. 2011. Assessing the value of earth observation for managing coral reefs: An example from the great barrier reef. *Science of the Total Environment* 409(21):4497–4503.

CCSP (U.S. Climate Change Science Program). 2009. *Best practice approaches for characterizing, communicating, and incorporating scientific uncertainty in climate decision making*. Washington, DC: National Oceanic and Atmospheric Administration.

Choy, S. L., R. O'Leary, and K. Mengersen. 2009. Elicitation by design in ecology: Using expert opinion to inform priors for Bayesian statistical models. *Ecology* 90(1):265–277.

Clemen, R. T. 1998. System models for decision making. In *System models for decision making*, edited by R. C. Dorf. Boca Raton, FL: CRC Press.

Clemen, R. T., and C. J. Lacke. 2001. Analysis of colorectal cancer screening regimens. *Health Care Management Science* 4(4):257–267.

Cooke, R. M., D. Nieboer, and J. Misiewicz. 2011. *Fat-tailed distributions: Data diagnostics and dependence*. Washington, DC: Resources for the Future.

Cox, L. A. T., Jr. 2012. Confronting deep uncertainties in risk analysis. *Risk Analysis* 32(10):1607–1629.

Cullen, A. C. 1994. Measures of compounding conservatism in probabilistic risk assessment. *Risk Analysis* 14(4):389–393.

DOE (U.S. Department of Energy). 2002. *Final environmental impact statement for a geologic repository for the disposal of spent nuclear fuel and high-level radioactive waste at Yucca Mountain, Nye County, Nevada (DOE/EIS-0250)*. Washington, DC: Department of Energy.

Duinker, P. N., and L. A. Greig. 2007. Scenario analysis in environmental impact assessment: Improving explorations of the future. *Environmental Impact Assessment Review* 27(3):206–219.

Ellig, J., and P. A. McLaughlin. 2012. The quality and use of regulatory analysis in 2008. *Risk Analysis* 32(5):855–880.

EPA (U.S. Environmental Protection Agency). 2003. Part III: Integrated summary and risk characterization for 2,3,7,8-tetrachlorodibenzo-p-dioxin (TCDD) and related compounds. In *Exposure and human health reassessment of 2,3,7,8-tetrachlorodibenzo-p-dioxin (tcdd) and related compounds. National Academies review draft*. Washington, DC: EPA.

EPA. 2011. *Expert Elicitation Task Force white paper*. Washington, DC: EPA. http://www.epa.gov/stpc/pdfs/ee-white-paper-final.pdf (accessed January 3, 2013).

Farber, D. A. 2007. Modeling climate change and its impacts: Law, policy, and science. *Texas Law Review* 86:1655.

FDA (U.S. Food and Drug Administration). 2003. *Quantitative assessment of relative risk to public health from foodborne Listeria monocytogenes among selected categories of ready-to-eat foods*. Washington, DC: U.S. Department of Health and Human Services and U.S. Department of Agriculture.

Feeny, D., W. Furlong, G. W. Torrance, C. H. Goldsmith, Z. Zhu, S. DePauw, M. Denton, and M. Boyle. 2002. Multiattribute and single-attribute utility functions for the Health Utilities Index Mark 3 system. *Medical Care* 40(2):113.

Flüeler, T. 2001. Options in radioactive waste management revisited: A proposed framework for robust decision making. *Risk Analysis* 21(4):787–800.

GAO (U.S. Governmental Accountability Office). 2006. *Human health risk assessment: EPA has taken steps to strengthen its process, but improvements needed in planning, data development, and training.* Washington, DC: Governmental Accountability Office.

Garabed, R., W. Johnson, J. Gill, A. Perez, and M. Thurmond. 2008. Exploration of associations between governance and economics and country level foot-and-mouth disease status by using Bayesian model averaging. *Journal of the Royal Statistical Society: Series A (Statistics in Society)* 171(3):699–722.

Gold, M. R. 1996. *Cost-effectiveness in health and medicine.* New York: Oxford University Press.

Gold, M. R., D. Stevenson, and D. G. Fryback. 2002. HALYs and QALYs and DALYs, oh my: Similarities and differences in summary measures of population health. *Annual Review of Public Health* 23(1):115–134.

Griffin, S., K. Claxton, and M. Sculpher. 2008. Decision analysis for resource allocation in health care. *Journal of Health Services Research and Policy* 13(Suppl 3):23–30.

Hahn, R., and P. Dudley. 2004. How well does the government do cost-benefit analysis? AEI-Brookings Joint Center Working Paper 04-01. https://www.law.upenn.edu/academics/institutes/regulation/papers/hahn_paper.pdf (accessed January 3, 2013).

———. 2007. How well does the government do cost-benefit analysis? *Review of Enivonmental Economics* 1(2):192–211.

Hahn, R. W., and P. C. Tetlock. 2007. Has economic analysis improved regulatory decisions? AEI-Brookings Joint Center Working Paper 07-08. http://papers.ssrn.com/sol3/papers.cfm?abstract_id=982233 (accessed January 3, 2012).

———. 2008. Has economic analysis improved regulatory decisions? *Journal of Economic Perspectives* 22(1):67–84.

Hora, S. 2007. Eliciting probabilities from experts. In *Advances in decision analysis: From foundations to applications,* edited by W. Edwards, R. F. Miles, and D. von Winterfeldt. Cambridge, UK: Cambridge University Press. Pp. 129–153.

Industrial Economics Incorporated. 2006. *Expanded expert judgment assessment of the concentration-response relationship between $PM_{2.5}$ exposure and mortality.* Research Triangle Park, NC: EPA.

Jarke, M., X. T. Bui, and J. M. Carroll. 1998. Scenario management: An interdisciplinary approach. *Requirements Engineering* 3(3):155–173.

Kandlikar, M., J. Risbey, and S. Dessai. 2005. Representing and communicating deep uncertainty in climate-change assessments. *Comptes Rendus Geosciences* 337(4):443–455.

Keeney, R. L., and H. Raiffa. 1976. *Decisions with multiple objectives.* Cambridge, UK: Cambridge University Press.

Kiker, G. A., T. S. Bridges, A. Varghese, T. P. Seager, and I. Linkov. 2005. Application of multicriteria decision analysis in environmental decision making. *Integrated Environmental Assessment and Management* 1(2):95–108.

Kuhnert, P. M. 2011. Four case studies in using expert opinion to inform priors. *Environmetrics* 22(5):662–674.

Kuhnert, P. M., T. G. Martin, and S. P. Griffiths. 2010. A guide to eliciting and using expert knowledge in Bayesian ecological models. *Ecology Letters* 13(7):900–914.

Leal, J., S. Wordsworth, R. Legood, and E. Blair. 2007. Eliciting expert opinion for economic models: An applied example. *Value in Health* 10(3):195–203.

Lempert, R. J. 2002. A new decision sciences for complex systems. *Proceedings of the National Academy of Sciences of the United States of America* 99(Suppl 3):7309–7313.

Lempert, R. J., and M. T. Collins. 2007. Managing the risk of uncertain threshold responses: Comparison of robust, optimum, and precautionary approaches. *Risk Analysis* 27(4):1009–1026.

Lempert, R. J., S. W. Popper, and S. C. Bankes. 2003. *Shaping the next one hundred years: New methods for quantitative, long-term policy analysis.* Santa Monica, CA: RAND.
Lempert, R. J., S. W. Popper, D. Groves, and S. C. Bankes. 2006. A general, analytic method for generating robust strategies and narrative scenarios. *Management Science* 52(4):514–528.
Levy, J. K., K. W. Hipel, and D. M. Kilgour. 2000. Using environmental indicators to quantify the robustness of policy alternatives to uncertainty. *Ecological Modelling* 130(1):79–86.
Levy, J., S. Chemerynski, and J. Tuchmann. 2006. Incorporating concepts of inequality and inequity into health benefits analysis. *International Journal for Equity in Health* 5(1):2.
Mahnovski, S. 2007. *Robust decisions and deep uncertainty: An application of real options to public and private investment in hydrogen and fuel cell technologies.* Santa Monica, CA: RAND.
Martin, T. G., M. A. Burgman, F. Fidler, P. M. Kuhnert, S. Low–Choy, M. McBride, and K. Mengersen. 2012. Eliciting expert knowledge in conservation science. *Conservation Biology* 26(1):29–38.
Merkhofer, M. W., and R. L. Keeney. 1987. A multiattribute utility analysis of alternative sites for the disposal of nuclear waste. *Risk Analysis* 7(2):173–194.
Morris, S., N. J. Devlin, and D. Parkin. 2007. *Economic analysis in health care.* Hoboken, NJ: Wiley.
NCI (National Cancer Institute). 2012. CISNet: Funding history and goals. http://cisnet.cancer.gov/about/history.html (accessed November 20, 2012).
Nichols, A. L., and R. J. Zeckhauser. 1988. The perils of prudence: How conservative risk assessments distort regulation. *Regulatory Toxicology and Pharmacology* 8(1):61–75.
NRC (National Research Council). 2001. *Arsenic in drinking water: 2001 update.* Washington, DC: National Academy Press.
———. 2006. *Health risks from dioxin and related compounds: Evaluation of the EPA reassessment.* Washington, DC: The National Academies Press.
———. 2009. *Science and decisions: Advancing risk assessment.* Washington, DC: The National Academies Press.
O'Leary, R. A., J. V. Murray, S. J. Low Choy, and K. L. Mengersen. 2008. Expert elicitation for Bayesian classification trees. *Journal of Applied Probability and Statistics* 3(1):95–106.
OMB (Office of Management and Budget). 2012. *Draft 2012 report to congress on the benefits and costs of federal regulations and unfunded mandates on state, local, and tribal entities.* Washington, DC: Office of Management and Budget.
Orme, M., J. Kerrigan, D. Tyas, N. Russell, and R. Nixon. 2007. The effect of disease, functional status, and relapses on the utility of people with multiple sclerosis in the UK. *Value in Health* 10(1):54–60.
Prato, T. 2003. Multiple-attribute evaluation of ecosystem management for the Missouri River system. *Ecological Economics* 45(2):297–309.
Raiffa, H. 1968. *Decision analysis—Introductory lectures on choices under uncertainty.* Reading, MA: Addison-Wesley.
Rhomberg, L.R. 2000. Dose-response analyses of the carcinogenic effects of trichloroethylene in experimental animals. *Environmental Health Perspectives* 108(Suppl 2):343–358.
Sloan, F. A., and C.-R. Hsieh. 2012. *Health economics.* Cambridge, MA: MIT Press.
Slottje, P., J. Van der Sluijs, and A. B. Knol. 2008. Expert elicitation: Methodological suggestions for its use in environmental health impact assessments. Letter report 630004001/2008. Utrecht: RIVM (National Institute for Public Health and the Environment). http://www.nusap.net/downloads/reports/Expert_Elicitation.pdf (accessed January 4, 2012).
Southwell, C., and D. von Winterfeldt. 2008. A decision analysis of options to rebuild the New Orleans flood control system. CREATE Research Archive. http://research.create.usc.edu/cgi/viewcontent.cgi?article=1052&context=published_papers (accessed January 4, 2013).

Steenland, K., G. Calvert, N. Ketchum, and J. Michalek. 2001. Dioxin and diabetes mellitus: An analysis of the combined NIOSH and Ranch Hand data. *Occupational and Environmental Medicine* 58(10):641-648.

U.S. Army Corps of Engineers. 1984. Lake Pontchartrain, Louisiana, and vicinity hurricane protection project: Reevaluation study. New Orleans, LA: U.S. Army Corps of Engineers. http://www.iwr.usace.army.mil/docs/hpdc/docs/19840700_Reevaluation_Study_Vol_1_Main_Report_Final_Sup_I_to_the_EIS.pdf (accessed January 4, 2013).

Viscusi, W., J. Hamilton, and C. Dockins. 1997. Conservative versus mean risk assessments: Implications for Superfund policies. *Journal of Environmental Economics and Management* 34(3):187-206.

Ye, M., K. F. Pohlmann, and J. B. Chapman. 2008. Expert elicitation of recharge model probabilities for the Death Valley regional flow system. *Journal of Hydrology* 354(1):102-115.

Appendix B

Committee Member Biographical Sketches

Frank A. Sloan, Ph.D. (*Chair*), is the J. Alexander McMahon Professor of Health Policy and Management and professor of economics at Duke University. Prior to joining the faculty at Duke in 1993, Dr. Sloan was a professor of economics at Vanderbilt University for nearly 10 years. Dr. Sloan's research interests are broad and include health care regulation and competition, health manpower, the cost effectiveness of medical technologies, aging and long-term care, and the social and economic costs of smoking and alcohol abuse. He has published more than 20 books and 300 journal articles and book chapters and has served on several national advisory councils and committees. He is currently on the editorial boards of the journals *Applied Health Economics and Health Policy* and *Journal of American Health Policy*. Dr. Sloan was elected a member of the Institute of Medicine (IOM) in 1982. He has been a member of several IOM committees and served as chair of both the IOM Committee to Evaluate Cancer in Low- and Middle-Income Countries (December 2004 to December 2006) and the IOM Committee to Evaluate Vaccine Purchase and Finance in the United States (January 2002 to August 2003). Dr. Sloan received his Ph.D. in economics from Harvard University.

James S. Hoyte, J.D., is a lecturer on environmental science and public policy at the Kennedy School of Government and assistant to the president and associate vice president of Harvard University. Mr. Hoyte is a specialist in environmental justice. He is currently co-program director of the Harvard Working Group on Environmental Justice, which brings together Harvard faculty from many disciplines to examine issues of environmental justice within the United States and around the world. Mr. Hoyte served as Massachusetts secretary of environmental affairs from 1983 to 1988. In that role he was responsible for oversight of the planning and management of environmental and natural resource conservation policies and programs for the Commonwealth of Massachusetts. As secretary, he was also founding chairman of the Board of the Massachusetts Water Resources Authority and oversaw management of the Boston Harbor Clean-up Project. Mr. Hoyte serves on the boards of directors of several environmental organizations, including the Union of Concerned Scientists, the Trust for Public Land, and the Massachusetts Environmental Trust. Mr. Hoyte received his J.D. from Harvard Law School.

Roger E. Kasperson, Ph.D., is a research professor and distinguished scientist in the Graduate School of Geography at Clark University. Before joining the faculty at Clark, Dr. Kasperson taught at the University of Connecticut and Michigan State University. He has published widely in the areas of risk analysis, risk communication, global environmental change, risk and ethics, and environmental policy. Dr. Kasperson was elected a member of the National Academy of Sciences in 2003 and the American Academy of Arts and Sciences in 2004. He has been a consultant or advisor to numerous public and private agencies on energy and environmental issues and has served on various committees of the National Research Council and the Council of the Society for Risk Analysis. From 1992 to 1996 he chaired the International Geographical Union Commission on Critical Situations/Regions in Environmental Change. He was vice president for academic affairs at Clark University from 1993 to 1996, and in 1999 he was elected director of the Stockholm Environment Institute, a post he held through 2004. He now serves on the Human Dimensions of Global Change Committee and the Committee on Strategic Advice for the Climate Change Program of the National Research Council. Dr. Kasperson has a Ph.D. in geography from the University of Chicago.

Emmett B. Keeler, Ph.D., is a professor in the Pardee RAND Graduate School and an adjunct professor at the University of California, Los Angeles, Public Health School, where he has taught cost effectiveness and decision analysis for many years. He led the multisite Improving Chronic Illness Care Evaluation. He analyzed health outcomes and episodes of spending for the RAND Health Insurance Experiment. He has worked on the theory and practice of decision analysis and cost-effectiveness analysis of clinical procedures and cancer screening. An elected member of the Institute of Medicine (IOM), he has participated in IOM committees on the science of lie detection, the economic costs of lack of insurance, the use of health measures in regulatory analysis, national health accounts, and geographic variation in health care spending. Dr. Keeler has a Ph.D. in mathematics from Harvard University.

Sarah B. Kotchian, Ph.D., Ed.M., M.P.H., was the associate director for planning for the Institute for Public Health at the University of New Mexico (UNM) until her retirement in 2007. In conjunction with the Centers for Disease Control and Prevention, Dr. Kotchian worked to strengthen national, state, and local environmental health services and to promote environmental health leadership. Before joining the faculty at UNM, she was the director of the Albuquerque Environmental Health Department for more than 14 years. Under Dr. Kotchian's direction, the Albuquerque Environmental Health Department administered comprehensive city- and county-wide programs in the areas of air quality, environmental health planning, food protection, noise control, hazardous waste, pollution prevention, groundwater quality and protection, landfill characterization and remediation, integrated vector management, vehicle emissions, epidemiology, geographic information systems, public information, and animal services. She is a recognized leader, author, and speaker on the formation of local, state, and national environmental and public health policy, leadership, and practice. She is the recipient of several awards, including the National Environmental Health Association's Walter Mangold Award. She served on the executive board of the American Public Health Association (APHA) and chaired its Subcommittee on Environment and Health. She was chair of the National Conference of Local Environmental Health Administrators. Other past activities include service on the National Environmental Health Science and Protection Accreditation Council; service on the Council on Education for Public Health, the national accrediting body for graduate public health education; and chair of the APHA Section on Environment. Dr. Kotchian has also served on a number of state and national committees on the environment and public health. She received a Ph.D. from UNM, an Ed.M. from Harvard University, and an M.P.H. from the University of Washington.

Joseph V. Rodricks, Ph.D., is a principal of ENVIRON International, a technical consulting firm, and a visiting professor at the Johns Hopkins University Bloomberg School of Public Health. ENVIRON was founded in 1982 by Dr. Rodricks and four associates and now has more than 900 employees in 13 countries. Dr. Rodricks came to consulting after a 15-year career as a scientist at the U.S. Food and Drug Administration (FDA). He entered the agency's Bureau of Science after receiving degrees in chemistry (Massachusetts Institute of Technology) and biochemistry (University of Maryland). He spent 7 years as a laboratory scientist investigating the chemistry, metabolism, and toxicology of aflatoxins and other natural toxins; during this same period he undertook a year of postdoctoral work at Berkeley, where he pursued studies of paralytic shellfish poison. The remainder of his FDA career was devoted to the development and application of quantitative risk assessment methods, and his professional life has continued to center on these subjects. He was the FDA's deputy associate commissioner for health affairs from 1977 until he resigned in 1980. During his last 4 years at the agency he was heavily involved in the formation of the National Toxicology Program and in a host of other interagency efforts, including the development of the first federal guidelines for the conduct of risk assessment. He has provided consulting services for manufacturers, government agencies, and the World Health Organization. His experience extends from pharmaceuticals, medical devices, and foods to occupational chemicals and environmental contaminants. He currently serves on the National Research Council (NRC) Board on Environmental Studies and Toxicology and has served on 24 committees of the NRC and the Institute of Medicine, including the committee that produced the seminal work *Risk Assessment in the Federal Government* (1983). Dr. Rodricks has received distinguished service awards from the Society for Risk Analysis and from the Food and Drug Law Institute. In 2003 he was awarded a lifetime appointment as a national associate of the National Academy of Sciences. His best-selling book, *Calculated Risks*, published by Cambridge University Press and recently released in a fully revised second edition, was given an award by the American Medical Writer's Association.

Susan L. Santos, Ph.D., M.S., is an assistant professor in the Department of Health Education and Behavioral Sciences at the University of Medicine and Dentistry of the New Jersey School of Public Health and holds a concurrent appointment to the VA War Related Illness and Injury Study Center in East Orange, New Jersey, where she serves as the associate director of education and risk communication. Dr. Santos is also the founder and principal of FOCUS GROUP, a consultancy specializing in risk communication, community relations, and health and environmental management. She combines her research and hands-on experience to aid federal, state, and local government agencies and private-sector clients with the design, implementation, and evaluation of health, safety, and environmental risk communication and community involvement programs. Prior to forming FOCUS GROUP, Dr. Santos served as director of corporate risk assessment services for ABB Environmental. She also worked for 8 years for EPA Region 1 in the areas of hazardous waste management and served as the agency lead for risk assessments. Dr. Santos is currently involved in implementing community engagement strategies for the cleanup of hazardous waste sites and conducting public health risk and crisis communication training. Her research interests include exploring how to communicate the results of health studies to community members, including to low-literacy audiences, and methods for evaluating stakeholder-involvement programs. She is a member of the Society for Risk Analysis and the Society for Public Health Education and has served on previous National Academy of Sciences committees. Dr. Santos has a Ph.D. in law, policy, and society from Northeastern University and an M.S. in civil engineering and public health from Tufts University.

Stephen H. Schneider, Ph.D., M.S.,[1] the Melvin and Joan Lane Professor for Interdisciplinary Environmental Studies, was a professor of biological sciences and professor by courtesy in the Department of Civil Engineering at Stanford University beginning in September 1992. He was a senior fellow in the Woods Institute for the Environment. In 1975 he founded the interdisciplinary journal *Climatic Change* and served as its editor. Dr. Schneider was elected a member of the National Academy of Sciences in 2002. He served on numerous National Research Council committees, including the Committee on the Human Dimensions of Global Change and the Institute of Medicine Committee on Decision Making Under Uncertainty. He was a coordinating lead author in Working Group II of the Intergovernmental Panel on Climate Change (IPCC) beginning in 1997 and was a lead author in Working Group I from 1994 to 1996. He was also a lead author of the IPCC guidance paper on uncertainties. He was a member of the California Climate Change Advisory Committee to advise the governor and state agencies on climate change policy. Dr. Schneider received both the National Conservation Achievement Award from the National Wildlife Federation and the Edward T. Law Roe Award of the Society of Conservation Biology in 2003. Dr. Schneider's global change research interests included climatic change, climatic modeling, global warming, the ecological and economic implications of climatic change, integrated assessment of global change policy, uncertainties, dangerous anthropogenic interference with the climate system, and abrupt climate change. Dr. Schneider received his Ph.D. in 1971 in mechanical engineering and plasma physics from Columbia University.

Stephanie Tai, J.D., Ph.D., is an assistant professor of law at the University of Wisconsin School of Law, where she teaches courses in administrative law, environmental law, property, environmental justice, risk regulation, and comparative Asian environmental law. Prior to joining the faculty at University of Wisconsin, she taught at Georgetown Law Center. She also has worked as the editor-in-chief of the *International Review for Environmental Strategies*, a publication of the Institute for Global Environmental Strategies in Japan. She served as a judicial law clerk to the Hon. Ronald

---
[1] Deceased, July 2010.

Lee Gilman on the U.S. Court of Appeals for the Sixth Circuit and then worked as an appellate attorney in the Environment and Natural Resources Division of the U.S. Department of Justice, during which time she briefed and argued cases involving a range of issues, from the protection of endangered cave species in Texas to the issuance of dredge-and-fill permits under the Clean Water Act. Dr. Tai's research interests include the interactions between environmental and health sciences and administrative law. She has written on the consideration of scientific studies and environmental justice concerns by administrative agencies and is currently studying the role of scientific dialogues before the judicial system. Dr. Tai received her J.D. from Georgetown University Law Center and her Ph.D. from Tufts University.

Detlof von Winterfeldt, Ph.D., is a professor of industrial and systems engineering at the Viterbi School of Engineering of the University of Southern California (USC) with a joint appointment as professor of public policy at the USC Sol Price School of Public Policy. From 2009 to early 2012 he served as director of the International Institute for Applied Systems Analysis (IIASA) in Austria. Concurrently with his IIASA appointment he was a Centennial Professor of Management Science at the London School of Economics and Political Science. In 2004 he co-founded the National Center for Risk and Economic Analysis of Terrorism Events, the first university-based center of excellence funded by the U.S. Department of Homeland Security, serving as its director until 2008. His research interests concern the foundation and practice of decision and risk analysis as applied to the areas of technology development, environmental risks, natural hazards, and terrorism. He has served on many committees and panels of the National Science Foundation and the National Academy of Sciences (NAS), including the NAS Board on Mathematical Sciences and their Applications. He is an elected fellow of the Institute for Operations Research and the Management Sciences (INFORMS) and of the Society for Risk Analysis. He has received the Ramsey Medal for distinguished contributions to decision analysis from the Decision Analysis Society of INFORMS and the Gold Medal for advancing the field from the International Society for Multicriteria Decision Making. He received his Ph.D. in mathematical psychology from the University of Michigan.

Robert B, Wallace, M.D., M.Sc., is the Irene Ensminger Stecher Professor of Epidemiology and Internal Medicine at the University of Iowa Colleges of Public Health and Medicine. He was formerly head of the department of preventive medicine at the University of Iowa College of Medicine and director of the University of Iowa Cancer Center. Dr. Wallace's research interests include cancer epidemiology and prevention; the causes and prevention of chronic, disabling diseases among older persons; women's health issues; and risk factors for cardiovascular disease. He is a principal investigator of several research projects on the health of older persons. He received his M.D. from the Northwestern University School of Medicine. He was elected a member of the Institute of Medicine (IOM) in 2001 and is currently chair of IOM's Board on Military and Veterans Health.

Appendix C

Meeting Agendas

The National Academies
Institute of Medicine
Committee on Decision Making Under Uncertainty

Meeting One
Thursday, November 1, 2007

Meeting Location:
The National Academies Keck Center
Room 110
500 Fifth Street, NW
Washington, DC 20001

1:30–3:00 p.m.   Discussion of Charge

　　　　　　　　George Gray, Ph.D.
　　　　　　　　Assistant Administrator, Office of Research and Development, EPA

3:00–3:30 p.m.   **Adjourn Open Session**

Meeting Two
Thursday, January 10, 2008

Meeting Location:
The National Academies Keck Center
Room 110
500 Fifth Street, NW
Washington, DC 20001

12:30–1:30 p.m.	Uncertainty Analysis at the EPA

        Michael Callahan
        Formerly: EPA, ORD Regional Science Liaison, Region 6

1:30–2:45 p.m.	*Robert Brenner*
        Director, Office of Policy Analysis and Review
        Office of Air and Radiation

        Bryan Hubbell
        Office of Air Quality Planning and Standards
        Innovative Strategies and Economics Group

2:45 p.m.	Adjourn Open Session

Friday, January 11, 2008

8:30–9:00 a.m.	Incorporation of Risk and Uncertainty Analysis Principles into Decision Making Within the Army Corps of Engineers

        Todd S. Bridges, Ph.D.
        Senior Scientist, Environmental Sciences
        U.S. Army Engineer Research and Development Center

9:00–9:30 a.m.	Discussion with the Committee

9:30–10:00 a.m.	Stakeholder Views About the Use of Probabilistic Risk Assessments at EPA

        Carol J. Henry, Ph.D., D.A.B.T.
        Retired, American Chemistry Council

APPENDIX C   257

      John Balbus, M.D., M.P.H.
      Chief Health Scientist, Program Director
      Environmental Defense Fund

10:00–10:30 a.m. Discussion with Panelists

10:30 a.m.   Adjourn Open Session

Meeting Three
Friday, February 15, 2008

Meeting Location:
The National Academies Keck Center
Room 100
500 Fifth Street, NW
Washington, DC 20001

9:00–9:40 a.m.   A Brief History of Uncertainty Analysis and Decision Making at the EPA

         Granger Morgan, Ph.D.
         Carnegie Mellon University

9:40–10:30 a.m. Uncertainty and Decision Making at the FDA

         Robert Buchanan, Ph.D.
         Director, Office of Science
         Center for Food Safety and Applied Nutrition
         Food and Drug Administration

10:30 a.m.–   Uncertainty and Decision Making: Challenges at the
12:00 p.m.   EPA

         Marianne Horinko
         The Horinko Group
         Formerly, Assistant Administrator for Solid Waste and Emergency Response, EPA

         Lynn Goldman, M.D.
         Johns Hopkins University Bloomberg School of Public Health
         Former, Assistant Administrator for Office of Prevention, Pesticides and Toxic Substances, EPA

	Dan Greenbaum Health Effects Institute
12:00–12:30 p.m.	**Discussion**
12:30–1:30 p.m.	**Break for Lunch**
1:30–3:30 p.m.	**Community Involvement in EPA Decision Making Under Uncertainty** *Peggy Shepard* Executive Director, West Harlem Environmental Action, Inc. *Nicky Sheats, Ph.D., J.D.* Director, Center for the Urban Environment John S. Watson Institute for Public Policy Thomas Edison State College *Julia Brody, Ph.D.* Executive Director, Silent Spring Institute
3:30–4:30 p.m.	**Discussion with Panelists**
4:30 p.m.	**Adjourn**